NOISE REDUCTION TECHNIQUES
IN ELECTRONIC SYSTEMS

NOISE REDUCTION TECHNIQUES IN ELECTRONIC SYSTEMS

HENRY W. OTT
Member of Technical Staff
Bell Laboratories

A Wiley-Interscience Publication

JOHN WILEY & SONS   New York • London • Sydney • Toronto

Library of Congress Cataloging in Publication Data

Ott, Henry W   1936–
  Noise reduction techniques in electronic systems.

  "A Wiley-Interscience publication."
  Includes bibliographies and index.
  1. Electronic circuits—Noise. I. Title.

TK7867.5.087   1976   621.3815'3   75-33165
ISBN 0-471-65726-3

PRINTED IN THE UNITED STATES OF AMERICA

10 9 8 7 6 5

Kathleen

PREFACE

This book covers the practical aspects of noise suppression and control in electronic circuits. It is intended primarily for the practicing engineer who is involved in the design of electronic equipment or systems, and also as a text for teaching the practical aspects of noise suppression. The concepts of noise reduction presented in the book can be applied to circuits operating from audio frequencies through VHF. More emphasis is placed on low- to mid-frequency noise problems, however, since these are the least documented in existing literature.

Some of the most difficult and frustrating problems faced by design engineers concern elimination of noise from their circuits or systems. Most engineers are not well equipped to handle noise problems, since the subject is not normally taught in engineering schools, and what literature is available is widely scattered among many different journals.

Solutions to noise problems are usually found by trial and error with little or no understanding of the mechanisms involved. Such efforts are very time consuming and the solutions may prove unsatisfactory if the equipment is moved to a new environment. This situation is unfortunate, since most of the principles involved are simple and can be explained by elementary physics.

This text began as a set of lecture notes for an out-of-hours course given at Bell Laboratories and later presented as part of the in-hours continuing education program at the laboratories. The approach used in the text is design oriented, with the amount and complexity of mathematics kept to a minimum. In some cases, models representing physical phenomena have been simplified to provide more useful results. By making realistic simplifying assumptions, results having clear physical meaning are obtained.

The organization of the material is as follows. Chapter 1 is an introduction to the subject of noise reduction. Chapters 2 and 3 cover the two primary means of noise control: shielding and grounding, respectively. Chapter 4 covers other noise reduction techniques such as balancing, decoupling, and filtering. Chapter 5, on passive components, covers the characteristics that affect the components' noise performance and their use in noise reduction circuitry. Chapter 6 provides a detailed analysis of the shielding effectiveness of metallic sheets. Chapter 7 covers relays and switches and discusses methods of reducing noise generated by these devices. Chapter 8 covers intrinsic noise sources that result in a theoretical minimum level of noise present in a circuit. Chapter 9 discusses noise in transistors and integrated circuits.

At the end of each chapter is a summary of the most important points discussed. For those desiring additional information a bibliography is also included. In addition, Appendix A discusses the decibel and its use in noise measurements on voice-frequency analog communications systems. Appendix B (presented in the form of a check list) is an overall summary of the more commonly used noise reduction techniques. Review problems for each chapter can be found in Appendix D with answers in Appendix E.

I wish to express my gratitude to Mr. S. D. Williams, Jr., who collaborated with me on an original set of notes for a noise-control seminar. That work provided the seed from which this book grew. I am also grateful to the many students whose enthusiasm provided the incentive to continue this work. Special thanks to Mr. F. P. Sullivan and Miss A. L. Wasser for their technical editing of the manuscript, and to Mr. L. E. Morris and Mr. D. N. Heirman for their many helpful suggestions. In addition I would like to thank all my colleagues who reviewed the manuscript, for their useful comments. Finally, I would like to express my gratitude to Bell Laboratories for their cooperation and support.

HENRY W. OTT

Whippany, New Jersey
July 1975

CONTENTS

SYMBOLS

A	Area
A	Voltage gain
A	Absorption loss (dB)
B	Magnetic flux density
B	Multiple reflection correction factor (dB)
B	Noise bandwidth
C	Capacitance
C_T	Distributed capacitance of transmission line
c	Center to center distance
D	Distance between conductors
d	Diameter
E	Electric field
e	Base of natural logarithm (2.7183)
F	Noise factor
f	Frequency
f_c	Shield cutoff frequency
f_0	3-dB bandwidth
f_r	Resonant frequency
f_α	Alpha cutoff frequency
G	Available power gain
G_s	Source admittance
g_{fs}	Forward transconductance
g_{11}	Input admittance
H	Magnetic field
h	Height of conductor above ground plane.
h	Spacing of flat conductors.
I	Current
I_A	Minimum arcing current
I_{dc}	Direct current
I_f	$1/f$-noise current
I_G	Ground current
I_{gss}	Gate leakage current
I_L	Load Current
I_N	Arbitrary noise current

I_n	Equivalent input noise current
I_0	Current at time $t = 0$
I_s	Source current
I_S	Shield current
I_{sh}	Shot noise current
I_t	Thermal noise current
i	Instantaneous current
j	Unit vector along imaginary axis
K	Arbitrary integer
K	Constant
k	Boltzmann's constant
L	Inductance
L_c	Series inductance of capacitor
L_S	Shield inductance
L_T	Distributed inductance of transmission line
l	Length
M	Mutual inductance
m	Arbitrary integer
N	Network function
NF	Noise figure
P	Power
P_{no}	Noise power output
q	Charge on electron
R	Resistance
R	Reflection loss (dB)
R_{ac}	ac resistance
R_C	Conductor resistance
R_c	Series resistance of capacitor
R_{dc}	dc resistance
R_F	Resistance of fuse
R_G	Ground resistance
R_L	Load resistance
R_S	Shield resistance
R_s	Source resistance
R_{so}	Source resistance for minimum noise factor
R_T	Distributed resistance of transmission line
r	Radius
r_b'	Base resistance in T-equivalent transistor model
r_c	Collector resistance in T-equivalent transistor model
r_e	Emitter resistance in T-equivalent transistor model
S	Shielding effectiveness (dB)
S/N	Signal-to-noise power ratio

SNI	Signal-to-noise improvement factor
T	Temperature
T_e	Equivalent input noise temperature
T_0	Standard reference temperature
t	Time
t	Thickness
t_r	Pulse rise time
V	Voltage
V_A	Minimum arcing voltage
V_B	Glow discharge breakdown voltage
V_C	Common mode (longitudinal) noise voltage
V_c	Contact voltage
V_{dc}	dc voltage
V_G	Ground voltage
V_G	Glow discharge sustaining voltage
V_L	Load voltage
V_M	Differential (metallic) noise voltage
V_N	Arbitrary noise voltage
V_n	Equivalent input noise voltage
V_{nd}	Equivalent input device noise voltage
V_{no}	Output noise voltage
V_{nt}	Total equivalent input noise voltage
V_s	Source voltage
V_S	Shield voltage
V_t	Thermal noise voltage
w	Width of conductor
Z	Impedance
Z_c	Collector impedance
Z_e	Emitter impedance
Z_o	Characteristic impedance of transmission line
Z_0	Characteristic impedance of medium
Z_s	Shield impedance
Z_w	Wave impedance
α	Common base current gain
β	Common emitter current gain
γ	Correlation coefficient
δ	Skin depth
ϵ	Dielectric constant
ϵ_r	Relative dielectric constant
ζ	Damping factor
η	Shield factor
θ	Angle

λ	Wavelength
μ	Permeability
μ_r	Relative permeability
π	3.1416
ρ	Resistivity
ρ_r	Relative resistivity
σ	Conductivity
σ_r	Relative conductivity
ϕ	Magnetic flux
ω	Radian frequency ($2\pi f$)
ω_c	$2\pi f_c$

NOISE REDUCTION TECHNIQUES IN ELECTRONIC SYSTEMS

1 THE INTERFERENCE PROBLEM

Widespread use of electric and electronic circuits for communication, power distribution, automation, computation, and other purposes makes it necessary for diverse circuits to operate in close proximity. All too often these circuits affect each other adversely. Electromagnetic interference (EMI)* has become a major problem for circuit designers, and it is likely to become more severe in the future. The large number of electronic devices in common use is partly responsible for this trend. In addition, the use of integrated circuits is reducing the size of electronic equipment. As circuitry becomes smaller and more sophisticated, more circuits are crowded into less space, thus increasing the probability of interference.

Today's equipment designers need to do more than just make their circuits operate under ideal conditions in the laboratory. Besides that obvious task, they must also make sure the equipment will work in the "real world," with other equipment nearby. This means the equipment should not be affected by external noise sources, and should not itself be a source of noise. Elimination—or really avoidance—of electromagnetic interference should be a major design objective.

In Fig. 1-1, the block diagram of a radio receiver is used as an example to depict the various types of interference that can occur in equipment. The wiring between stages conducts noise, and some stages radiate noise. In addition, ground currents from the various stages flow through a common ground impedance and produce a noise voltage on the ground bus. Electric and magnetic field coupling between signals in various conductors is also shown. These noise problems are examples of intraequipment interference that must be solved before the radio will operate in the laboratory. When the radio is installed in the "real world," it becomes exposed to additional external noise sources, such as shown in Fig. 1-2. Noise currents are conducted into the receiver on the ac power line, and the radio receiver is exposed to electromagnetic radiation from various sources. In this case the noise sources are not under the designer's control. However, the unit must still be designed to operate in this environment.

*Present usage favors the more general term EMI in place of the older radio frequency interference (RFI).

1

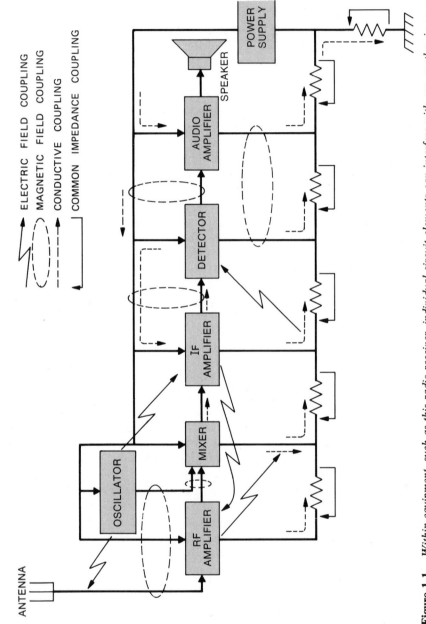

Figure 1-1. *Within equipment, such as this radio receiver, individual circuit elements can interfere with one another in several ways.*

2

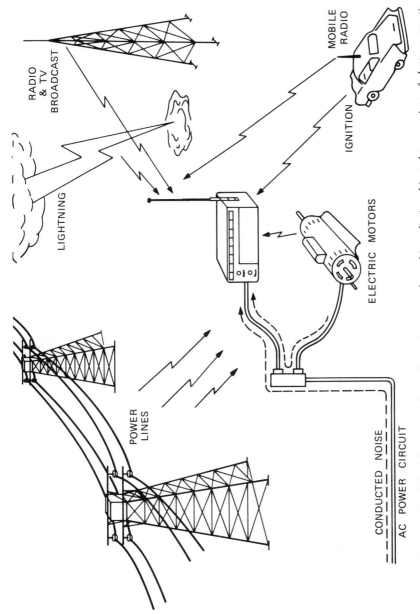
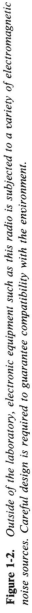

Figure 1-2. *Outside of the laboratory, electronic equipment such as this radio is subjected to a variety of electromagnetic noise sources. Careful design is required to guarantee compatibility with the environment.*

3

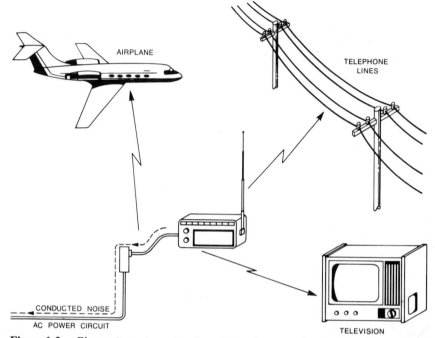

Figure 1-3. *Electronic equipment such as this radio can emit noise that may interfere with other circuits. Consideration of noise during equipment design can avoid these emissions.*

Figure 1-3 depicts the other side of the noise problem. The radio can be a source of noise that may interfere with other equipment. Parts of the circuit radiate noise directly, and the power cable conducts noise to other circuits. Noise current flowing in the power lead causes the lead to radiate additional noise. Designing equipment to minimize noise generation is equally as important as designing equipment that is not susceptible to interference.

DESIGNING FOR ELECTROMAGNETIC COMPATIBILITY

Electromagnetic compatibility (EMC) is the ability of equipment to function properly in its intended electromagnetic environment. EMC should be considered early in the design stages of a new piece of equipment. If the matter of EMC is ignored until a problem is revealed during testing, solutions are likely to be both unsatisfactory and expensive. As equipment development progresses from design, to testing, and to production, the variety of noise-elimination techniques available to the designer decreases

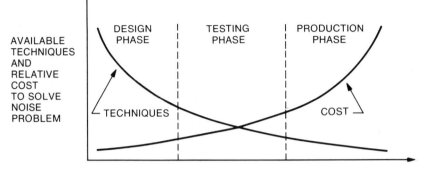

Figure 1-4. *As equipment development proceeds, the number of available noise-reduction techniques goes down. At the same time, the cost of noise reduction goes up.*

steadily. Concurrently, cost goes up. These trends are shown in Fig. 1-4. Early solutions to interference problems, therefore, are usually the best and the least expensive.

For example, if noise suppression is considered for one stage or subsystem at a time while the equipment is being designed, the noise reduction techniques are fairly simple and straightforward. Experience has shown that when noise suppression is handled this way, the designer should be able to produce equipment with 80–90% of the potential noise problems eliminated prior to testing.

On the other hand, a system designed with complete disregard to noise suppression will almost surely have noise problems when testing begins. Analysis at that time to find which of the many possible noise path combinations are contributing to the problem may not be at all simple or obvious. Solutions at this late stage usually involve the addition of extra components which are not integral parts of the circuit. Penalties paid include the added engineering cost and the cost of the components and their installation. There also may be size, weight, and power dissipation penalties.

Timely consideration should also be given to the task of minimizing the amount of noise generated by equipment, since the noise may interfere with other equipment. It is always desirable to control as much noise at the source as possible, since that approach can avoid an interference problem for a countless number of receiver circuits. To provide electromagnetic compatibility, therefore, equipment must be designed that does not adversely affect, and is not adversely affected by, any other equipment in the environment.

DEFINITIONS

Noise can be defined as any electrical signal present in a circuit other than the desired signal. An important exception to this definition is the distortion products produced in a circuit due to nonlinearities. These are really circuit design problems and not truly noise problems. Although these distortion products may be undesirable, they are not considered noise unless they get coupled into another part of the circuit. It follows from the definition of noise that a desired signal in one part of a circuit may be consider noise if inadvertently coupled into some other part of the circuit.

Noise sources can be grouped in three major categories. The first are the so-called intrinsic noise sources that arise from random fluctuations within physical systems. Examples of intrinsic noise are thermal and shot noise. Second are man-made noise sources, such as motors, switches, and transmitters. The third category is noise due to natural disturbances, such as lightning and sun spots.

Interference can be defined as the undesirable effect of noise. If a noise voltage causes unsatisfactory operation of a circuit, it is interference. Usually noise cannot be eliminated but only reduced in magnitude until it no longer causes interference.

Susceptibility is the capability of a device or circuit to respond to unwanted electrical energy (noise). The susceptibility level of a circuit or device is the noise environment in which the equipment can operate satisfactorily.

REGULATIONS

Some added insight into the problem of interference and the obligations of equipment designers can be gained from a review of the applicable government and military regulations and specifications.

In the United States, the Federal Communications Commission (FCC) regulates the use of radio and wire communications. Part of its responsibility concerns the control of interference. The FCC regulations, Part 15 (for radio-frequency devices) and Part 18 (for industrial, scientific, and medical equipment) both have sections covering the control of interference. These regulations specify the allowable amount of radiated energy for various classes of equipment.

A typical example of these regulations is Section 15.7, which covers general requirements for restricted radiation devices. The FCC defines a restricted radiation device as follows: "A device in which the generation of radio-frequency energy is intentionally incorporated into the design and in which the radio-frequency energy is conducted along wires or is radiated,..." This includes such items as garage door openers, 100-mW walkie-talkies, wireless microphones, and RF security devices. This section specifies that the total electromagnetic field produced at a distance of $\lambda/2\pi$ (approximately $\frac{1}{6}$ wavelength) from a device shall not exceed 15 μV/m unless licensed, and shall be operated with the minimum power possible to accomplish the desired purpose. Should harmful interference occur, prompt steps shall be taken to eliminate the interference. Should interference occur to essential communications or navigation services, the device must be shut down until the interference it causes has been corrected.

A second example of the FCC regulations can be found in Section 15.31, which covers any incidental radiation device defined as: "A device that radiates radio-frequency energy during the course of its operation although the device is not intentionally designed to generate radio-frequency energy." This includes switching circuits, transistors and control rectifiers, motors, power converters, automobile engines, and fluorescent lamps.

The FCC presently has only a very general noninterference requirement on incidental radiation devices. It states:

An incidental radiation device shall be operated so that the radio-frequency energy that is radiated does not cause harmful interference. In the event that harmful interference is caused, the operator of the device shall promptly take steps to eliminate the harmful interference.

The FCC has the authority to further regulate these devices with respect to interference.* To date, however, it has not acted on this authority relying on self regulation by industry. Should industry become lax in this respect, however, the FCC may move to exercise its jurisdiction in this area.

Devices that use radio waves for industrial, scientific, or medical purposes come under Part 18 of the FCC regulations. Included are medical diathermy equipment, industrial heating equipment, RF welders, devices used to produce physical changes in matter, and other related noncommunications devices. Part 18 specifies that the operator of industrial, scientific, or medical equipment that cause harmful interference to any authorized radio service shall promptly take steps to remedy the interference.

Another important source of information on interference regulations is available in military specifications. MIL-STD-461A, for example, sets limits on radiated interference at frequencies from 30 Hz to 10 GHz. Specific test methods and procedures for making the tests are contained in MIL-STD-462.

The categories of tests specified by MIL-STD-461A are organized as shown in the block diagram in Fig. 1-5. Tests are required for both

*Public Law 90-379, passed in 1968.

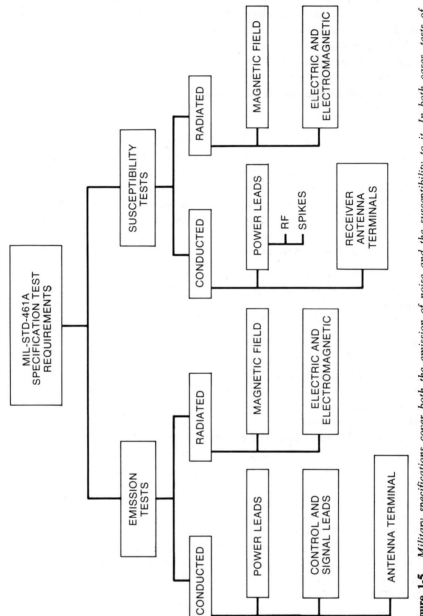

Figure 1-5. *Military specifications cover both the emission of noise and the susceptibility to it. In both cases, tests of conducted and radiated noise are specified.*

radiated and conducted emissions as well as radiated and conducted susceptibility. Notice that in addition to radiated emissions, the MIL specification gives the amount of noise that can be conducted out of equipment on power and signal leads, whereas the FCC regulations only specify radiated emissions. This MIL specification is a very comprehensive document and is often used by industry as a guideline for designing interference-free nonmilitary equipment.

The following examples of the requirements of MIL-STD-461A can also be used as reasonable criteria for the design of nonmilitary equipment.

Test RE01 requires that the magnetic field radiated emission at a distance of seven centimeters does not exceed the values shown in Fig. 1-6. Test RE02 requires that the broadband electric field radiated emission, at a distance of 1 m, does not exceed the values shown in Fig. 1-7.

The magnetic field radiated susceptibility requirement (Test RS01), requires that the test sample shall not exhibit any malfunction, degradation of performance, or deviation from specifications, when subjected to the magnetic field levels shown in Fig. 1-6. The electric field radiated suscepti- bility requirement (Test RS03, Notice 4), requires that the test sample shows no malfunction, degradation of performance, or deviation from specifications, when subjected to the electric field levels listed in Table 1-1.

Another test of general interest is the Spike Test (CS06) for conducted susceptibility on power leads. This test requires a 10-μs spike to be applied to all ungrounded dc or ac power leads. The spike must have an amplitude

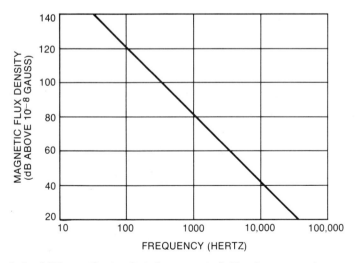

Figure 1-6. *MIL-specification limit for magnetic field radiation at a distance of 7 cm. The same curve also specifies the magnetic field susceptibility level. One gauss equals 10^{-4} tesla.*

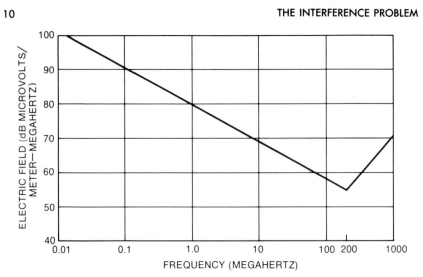

Figure 1-7. *MIL-specification limit for electric field radiation at a distance of 1 m.*

Table 1-1   Radiated Susceptibility

Frequency range (MHz)	Susceptibility level (V/m)
0.01– 1.9	1
2 – 29.9	5
30 –400	10

of twice the dc supply voltage or 100 V, whichever is smaller. Positive and negative, single and repetitive (6–10 pulses per second) spikes shall be applied for a period not to exceed 30 minutes. The spikes shall be synchronized with the test sample signal conditions such that they are most likely to produce interference. The test sample must not exhibit any malfunction, degradation of performance, or deviation from specifications, when the spike is applied. A requirement similar to this, possibly with a longer time duration, should be specified for most electronic equipment, since power-supply spikes are a major cause of interference and actual physical damage in solid-state circuits.

TYPICAL NOISE PATH

A block diagram of a typical noise path is shown in Fig. 1-8. As can be seen, three elements are necessary to produce a noise problem. First, there must be a noise source. Second, there must be a receiver circuit that is susceptible to the noise. Third, there must be a coupling channel to transmit the noise from the source to the receiver.

Figure 1-8. *Before noise can be a problem, there must be a noise source, a receiver that is susceptible to the noise, and a coupling channel that transmits the noise to the receiver.*

The first step in analyzing a noise problem is to define the problem. This is done by determining what the noise source is, what the receiver is, and how the source and receiver are coupled together. It follows that there are three ways to break the noise path: (1) the noise can be suppressed at the source, (2) the receiver can be made insensitive to the noise, or (3) the transmission through the coupling channel can be minimized. In some cases, noise suppression techniques must be applied to two or to all three parts of the noise path.

As an example, consider the circuit shown in Fig. 1-9. It shows a shielded dc motor connected to its motor-drive circuit. Motor noise is interfering with a low-level circuit in the same equipment. Commutator noise from the motor is conducted out of the shield on the leads going to the drive circuit. From the leads, noise is radiated to the low-level circuitry.

In this example, the noise source consists of the arcs between the brushes and the commutator. The coupling channel has two parts: conduction on the motor leads and radiation from the leads. The receiver is the low-level circuit. In this case, not much can be done about the source or the receiver. Therefore, the interference must be eliminated by breaking

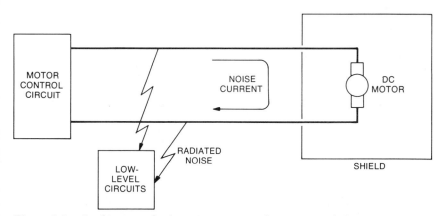

Figure 1-9. *In this example the noise source is the motor, and the receiver is the low-level circuit. The coupling channel consists of conduction on the motor supply leads and radiation from the leads.*

the coupling channel. Noise conduction out of the shield or radiation from the leads must be stopped, or both steps may be necessary. This example is discussed more fully in Chapter 5.

USE OF NETWORK THEORY

For the exact answer to the question of how any electric circuit behaves, Maxwell's equations must be solved. These equations are functions of three space variables (x, y, z) and of time (t). Solutions for any but the simplest problems are usually very complex. To avoid this complexity, an approximate analysis technique called "electric circuit analysis" is used during most design procedures.

Circuit analysis eliminates the spatial variables and provides approximate solutions as a function of time only. Circuit analysis assumes the following:

1. All electric fields are confined to the interiors of capacitors.
2. All magnetic fields are confined to the interiors of inductors.
3. Dimensions of the circuits are small compared to the wavelength(s) under consideration.

What is really implied is that external fields, even though actually present, can be neglected in the solution of the network. Yet, these external fields may not necessarily be neglected where their effect on other circuits is concerned.

For example, a 100-W power amplifier may radiate 100 mW of power.

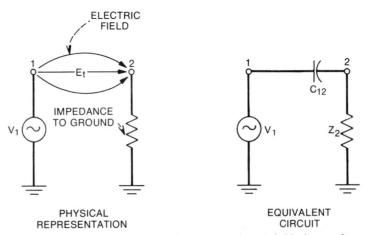

PHYSICAL
REPRESENTATION

EQUIVALENT
CIRCUIT

Figure 1-10. *When two circuits are coupled by an electric field, the coupling can be represented by a capacitor.*

These 100 mW are completely negligible as far as the analysis of the power amplifier is concerned. However, if only a small percentage of this radiated power is picked up on the input of a sensitive amplifier, it may produce a large noise signal.

Whenever possible, noise coupling channels are represented as equivalent lumped component networks. For instance, a time-varying electric field existing between two conductors can be represented by a capacitor connecting the two conductors (see Fig. 1-10). A time-varying magnetic field that couples two conductors can be represented by a mutual inductance between the two circuits (see Fig. 1-11).

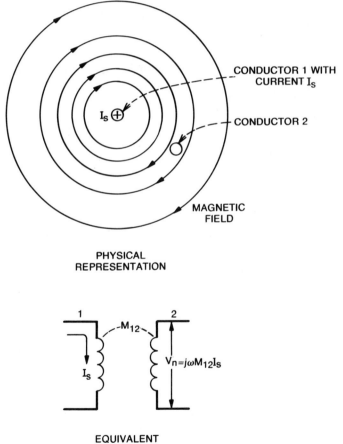

Figure 1-11. *When two circuits are coupled by a magnetic field, the coupling can be represented as a mutual inductance.*

For this approach to be valid, the physical dimensions of the circuits must be small compared to the wavelengths of the signals involved. This assumption is made throughout most of this book, and it is normally reasonable. For example, the wavelength of a 1-MHz signal is approximately 300 m. For a 300-MHz signal, it is 1 m. For most electronic circuits, the dimensions are smaller than this.

Even when the above assumption is not truly valid, the lumped component representation is still useful for the following reasons:

1. The solution of Maxwell's equations is not practical for most "real world" noise problems because of the complicated boundary conditions.

2. Lumped component representation, although it does not necessarily give the correct numerical answer, does clearly show how noise depends on the system parameters. On the other hand, the solution of Maxwell's equations, even if possible, does not show such dependence clearly.

In general the numerical values of the lumped components are extremely difficult to calculate with any precision, except for certain special geometries. One can conclude, however, that these components exist, and as will be shown, the results can be very useful even when the components are only defined in a qualitative sense.

METHODS OF NOISE COUPLING

Conductively Coupled Noise

One of the most obvious, but often overlooked, ways to couple noise into a circuit is on a conductor. A wire run through a noisy environment may pick up noise, and then conduct it to another circuit. There it causes interference. The solution is to prevent the wire from picking up the noise, or to remove the noise from it, by decoupling before it interferes with the susceptible circuit.

The major example in this category is noise conducted into a circuit on the power supply leads. If the designer of the circuit has no control over the power supply, or if other equipment is connected to the power supply, it becomes necessary to decouple the noise from the wires before they enter the circuit.

Coupling Through Common Impedance

Common impedance coupling occurs when currents from two different circuits flow through a common impedance. The voltage drop across the impedance seen by each circuit is influenced by the other. The classic example of this type of coupling is shown in Fig. 1-12. The ground currents

1 and 2 both flow through the common ground impedance. As far as circuit 1 is concerned, its ground potential is modulated by ground current 2 flowing in the common ground impedance. Some noise signal, therefore, is coupled from circuit 2 to circuit 1 through the common ground impedance.

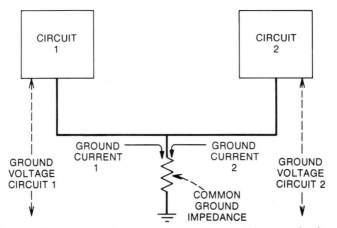

Figure 1-12. *When two circuits share a common ground, the ground voltage of each one is affected by the ground current of the other circuit.*

Another example of this problem is illustrated in the power distribution circuit shown in Fig. 1-13. Any change in the supply current required by circuit 2 will affect the voltage at the terminals of circuit 1, due to the

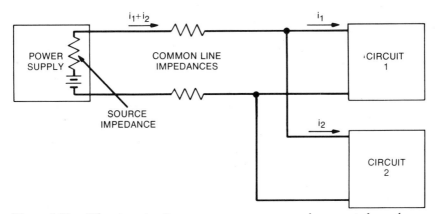

Figure 1-13. *When two circuits use a common power supply, current drawn by one circuit affects the voltage at the other circuit.*

common impedances of the power supply lines and the internal source impedance of the power supply. Some improvement can be obtained by connecting the leads from circuit 2 closer to the power supply output terminals, thus decreasing the magnitude of the common line impedance. The coupling through the power supply's internal impedance still remains, however.

Electric and Magnetic Fields

Radiated electric and magnetic fields provide another means of noise coupling. All circuit elements including conductors radiate electromagnetic fields whenever charge is moved. In addition to this unintentional radiation there is the problem of intentional radiation from sources such as broadcast stations and radar transmitters. When the receiver is close to the source (near field), electric and magnetic fields are considered separately. When the receiver is far from the source (far field), the radiation is considered as combined electric and magnetic or electromagnetic radiation.*

MISCELLANEOUS NOISE SOURCES

Galvanic Action

If dissimilar metals are used in the signal path in low-level circuitry, a noise voltage may appear due to the galvanic action between the two metals. The presence of moisture or water vapor in conjunction with the two metals produces a chemical wet cell. The voltage developed depends on the two metals used and is related to their positions in the galvanic series shown in Table 1-2. The farther apart the metals are on this table, the larger the developed voltage. If the metals are the same no potential difference can develop.

In addition to producing a noise voltage, the use of dissimilar metals can produce a corrosion problem. Galvanic corrosion causes positive ions from one metal to be transferred to the other one. This gradually causes the anode material to be destroyed. The rate of corrosion depends on the moisture content of the environment and how far apart the metals are in the galvanic series. The farther apart the metals are in the galvanic series, the faster the ion transfer. An undesirable, but common, combination of metals is aluminum and copper. With this combination, the aluminum is eventually eaten away. The reaction slows down considerably, however, if the copper is coated with lead–tin solder since aluminum and lead-tin solder are closer in the galvanic series.

*See Chapter 6 for an explanation of near field and far field.

Table 1-2   Galvanic Series

ANODIC END
(Most susceptible to corrosion)

Group I	1. Magnesium		13. Nickel (active)
			14. Brass
	2. Zinc		15. Copper
	3. Galvanized Steel		16. Bronze
Group II	4. Aluminum 2S	Group IV	17. Copper–Nickel Alloy
	5. Cadmium		18. Monel
	6. Aluminum 17ST		19. Silver Solder
			20. Nickel (passive)[a]
	7. Steel		21. Stainless Steel
	8. Iron		(passive)[a]
	9. Stainless Steel		
Group III	(active)		22. Silver
	10. Lead Tin Solder	Group V	23. Graphite
	11. Lead		24. Gold
	12. Tin		25. Platinum

CATHODIC END
(Least susceptible to corrosion)

[a]Passivation by immersion in a strongly oxidizing acidic solution.

Four elements are needed before galvanic action can occur:

1. Anode material (higher rank in Table 1-2)
2. Electrolyte (usually present as moisture)
3. Cathode material (lower rank in Table 1-2)
4. Conducting electrical connection between anode and cathode (usually present as a leakage path).

Galvanic action can take place even if moisture does not get between the anode and cathode. All that is needed is some moisture on the surface where the two metals come together, as shown in Fig. 1-14.

As seen in Table 1-2, the metals of the galvanic series are divided into five groups. When dissimilar metals must be combined, it is desirable to use metals from the same group.

Electrolytic Action

A second type of corrosion is due to electrolytic action. It is caused by a direct current flowing between two metals with an electrolyte (which could be slightly acidic ambient moisture) between them. This type of corrosion

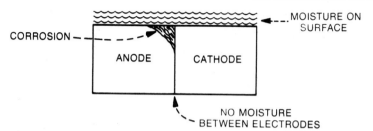

Figure 1-14. *Galvanic action can occur if two dissimilar metals are joined and moisture is present on the surface.*

does not depend on the two metals used and will occur even if both are the same. The rate of corrosion depends on the magnitude of the current and the conductivity of the electrolyte.

Triboelectric Effect

A charge can be produced on the dielectric material within a cable, if the dielectric does not maintain contact with the cable conductors. This is called the triboelectric effect. It is usually caused by mechanical bending of the cable. The charge acts as a noise voltage source within the cable. Eliminating sharp bends and cable motion minimizes this effect. A special "low noise" cable is available in which the cable is chemically treated to minimize the possibility of charge buildup on the dielectric.

Conductor Motion

If a wire is moved through a magnetic field, a voltage is induced between ends of the wire. Due to power wiring and other circuits with high current flow, stray magnetic fields exist in most environments. If a wire with a low-level signal is then allowed to move through this field, a noise voltage is induced in the wire. This problem can be especially troublesome in a vibrational environment. The solution is simple: prevent wiring motion with cable clamps and other tie-down devices.

METHODS OF ELIMINATING INTERFERENCE

The following chapters present techniques by which interference between electronic circuits can be eliminated, or at least reduced. The primary methods available for combatting interference are listed below:

1. Shielding
2. Grounding
3. Balancing

4. Filtering
5. Isolation
6. Separation and orientation
7. Circuit impedance level control
8. Cable design
9. Cancellation techniques (frequency or time domain).

Appendix B, presented in the form of a checklist, is a summary of the more commonly used noise reduction techniques. Even with all these methods available, it should be remembered that noise usually cannot be eliminated; it can only be minimized to the point where it no longer causes interference.

In all but the simplest cases, a single unique solution to the noise reduction problem may not exist. Compromises are generally required, and which of the many alternative solutions is the best can be the subject of considerable disagreement. In this book we will present the techniques which are useful for decreasing interference. Decisions on which techniques should be used in a specific case, however, are things that must be determined by the system design engineer.

SUMMARY

- Designing equipment that does not generate noise is as important as designing equipment that is not susceptible to noise.
- Noise suppression should be considered early in the design stage.
- Three items are necessary to produce a noise problem; there must be
    a noise source,
    a coupling channel,
    a susceptible receiver.
- The three primary means of noise coupling are
    conductive coupling,
    common impedance coupling,
    coupling by radiated electromagnetic fields.
- Metals in contact with each other, and in the signal path, must be galvanically compatible.
- A unique solution to most noise reduction problems does not always exist. There is usually more than one technique by which the noise objectives can be met.

BIBLIOGRAPHY

Bell Laboratories, *Physical Design of Electronic Systems*, Vol. 2, Chapter 5 (Electrochemistry and Protection of Surfaces), Prentice-Hall, Englewood Cliffs, N.J., 1970.

Cohen, T. J., and McCoy, L. G., "RFI—A New Look at an Old Problem," *QST*, March, 1975.

FCC Rules and Regulations, Vol. 2, U. S. Government Printing Office, Washington, D. C.

MIL-STD-461A, "Electromagnetic Interference Characteristic Requirements for Equipment," August, 1968.

MIL-STD-462, "Electromagnetic Interference Characteristics, Measurement of," July, 1967.

White, D. R. J., *Electromagnetic Interference and Compatibility*, Vol. 1 (Electrical Noise and EMI Specifications), Don White Consultants, 1971.

White, D. R. J., *Electromagnetic Interference and Compatibility*, Vol. 2 (EMI Test Methods and Procedures), Don White Consultants, 1974.

2 SHIELDING OF CONDUCTORS

The two primary ways to minimize unwanted noise pickup are shielding and grounding. This chapter is devoted to the subject of shielding, and Chapter 3 covers grounding. The techniques of shielding and grounding are closely interrelated, however, and these two chapters should be studied together as a single unit. In this chapter, for example, it is shown that a cable shield used to suppress electric fields should be grounded, but Chapter 3 explains where that ground should be made.

When properly used, shields can reduce the amount of noise coupling considerably. Shields may be placed around components, circuits, complete assemblies, or cables and transmission lines. This chapter is concerned with the shielding of conductors, although the same basic principles apply to any type of shielding. Chapter 6 contains additional information on other types of shielding. In this chapter, the following three assumptions are made:

1. Shields are made of nonmagnetic materials, and have a thickness much less than a skin depth at the frequency of interest.*
2. The receiver is not coupled so tightly to the source that it loads down the source.
3. Induced currents in the receiver signal circuit are small enough not to distort the original field. (This does not apply to a shield around the receiver circuit.)

To permit the problem of shielding to be studied, we shall represent the coupling between two circuits by lumped capacitance and inductance between the conductors. The circuit can then be analyzed by normal network theory.

Three types of coupling are considered. The first is capacitive, or electric, coupling, which is due to the interaction of electric fields between circuits. This type of coupling is commonly identified in the literature as electrostatic coupling, an obvious misnomer since the fields are not static.

*If the shield is thicker than a skin depth, some additional shielding is present besides that calculated by methods in this chapter. This effect is discussed further in Chapter 6.

The second is inductive, or magnetic, coupling, which results from the interaction between the magnetic fields of two circuits. This type of coupling is commonly described as electromagnetic, again misleading terminology since no electric fields are involved. The third is a combination of electric and magnetic fields and is appropriately called electromagnetic coupling or radiation. The techniques developed to cope with electric coupling and magnetic coupling will normally, when used together, be appropriate for the electromagnetic case. For analysis in the near field, we normally consider the electric and magnetic fields separately, whereas the electromagnetic field case is considered when the problem is in the far field.* The circuit which is causing the interference is called the source, and the circuit being affected by the interference is called the receiver.

CAPACITIVE COUPLING

A simple representation of capacitive coupling between two conductors is shown in Fig. 2-1. Capacitance C_{12} is the stray capacitance between conductors 1 and 2. Capacitance C_{1G} is the capacitance between conductor 1 and ground, C_{2G} is the total capacitance between conductor 2 and ground, and R is the resistance of circuit 2 to ground. The resistance R results from the circuitry connected to conductor 2 and is not a stray component. Capacitance C_{2G} consists of both the stray capacitance of conductor 2 to ground and the effect of any circuitry connected to conductor 2.

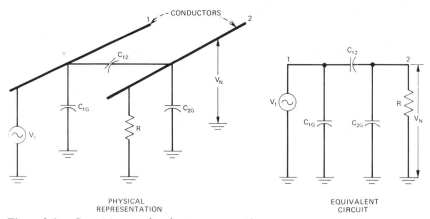

PHYSICAL
REPRESENTATION

EQUIVALENT
CIRCUIT

Figure 2-1. *Capacitive coupling between two conductors.*

*See Chapter 6 for definitions of near and far fields.

The equivalent circuit of the coupling is also shown in Fig. 2-1. Consider the voltage V_1 on conductor 1 as the source of interference and conductor 2 as the affected circuit or receiver. Any capacitance connected directly across the source, such as C_{1G} in Fig. 2-1, can be neglected since it has no effect on the noise coupling. The noise voltage V_N produced between conductor 2 and ground can be expressed as follows:

$$V_N = \frac{j\omega[C_{12}/(C_{12}+C_{2G})]}{j\omega + 1/R(C_{12}+C_{2G})} V_1. \tag{2-1}$$

Equation 2-1 does not show clearly how the pickup voltage depends on the various parameters. Equation 2-1 can be simplified for the case when R is a lower impedance than the impedance of the stray capacitance C_{12} plus C_{2G}. In most practical cases this will be true. Therefore, for

$$R \ll \frac{1}{j\omega(C_{12}+C_{2G})},$$

Equation 2-1 can be reduced to the following:

$$\boxed{V_N = j\omega R C_{12} V_1.} \tag{2-2}$$

This is the most important equation describing the capacitive coupling between two conductors, and it clearly shows how the pickup voltage depends on the parameters. Equation 2-2 shows that the noise voltage is directly proportional to the frequency ($\omega = 2\pi f$) of the noise source, the resistance R of the affected circuit to ground, the capacitance C_{12} between conductors 1 and 2, and the magnitude of the voltage V_1.

Assuming that the voltage and frequency of the noise source cannot be changed, this leaves only two remaining parameters for reducing capacitive coupling. The receiver circuit can be operated at a lower resistance level, or capacitance C_{12} can be decreased. Capacitance C_{12} can be decreased by proper orientation of the conductors, by shielding (described in the next section), or by physically separating the conductors. If the conductors are moved farther apart, C_{12} decreases, thus decreasing the induced voltage on conductor 2.* The effect of conductor spacing on capacitive coupling is shown in Fig. 2-2. As a reference, 0 dB is the coupling when the conductors are separated by three times the conductor diameter. As can be seen in the figure, little additional attenuation is gained by spacing the conductors

---

*The capacitance between two parallel conductors of diameter d and spaced D apart is $C_{12} = \pi\epsilon/\cosh^{-1}(D/d)$, (F/m). For $D/d > 3$, this reduces to $C_{12} = \pi\epsilon/\ln(2D/d)$, (F/m), where $\epsilon = 8.85 \times 10^{-12}$ farads per meter (F/m) for free space.

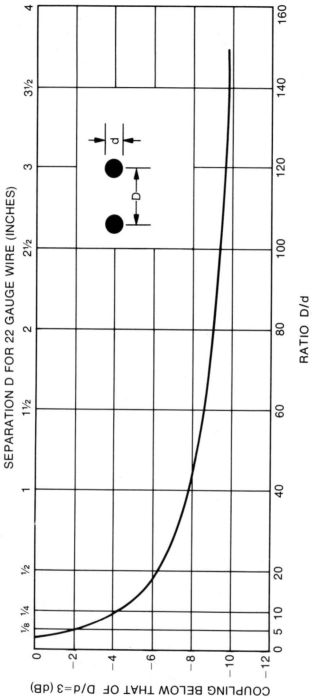

Figure 2-2. *Effect of conductor spacing on capacitive coupling. In the case of 22 gauge wire, most of the attenuation occurs in the first inch of separation.*

24

a distance greater than 40 times their diameter (1 in. in the case of 22-gauge wire).

If the resistance from conductor 2 to ground is large, such that

$$R \gg \frac{1}{j\omega(C_{12}+C_{2G})} \, ,$$

then Eq. 2-1 reduces to

$$V_N = \left(\frac{C_{12}}{C_{12}+C_{2G}} \right) V_1. \qquad (2\text{-}3)$$

Under this condition, the noise voltage produced between conductor 2 and ground is due to the capacitive voltage divider C_{12} and C_{2G}. The noise voltage is independent of frequency and is of a larger magnitude than when R is small.

A plot of Eq. 2-1 versus ω is shown in Fig. 2-3. As can be seen, the maximum noise coupling is given by Eq. 2-3. The figure also shows that the actual noise voltage is always less than or equal to the value given by Eq. 2-2. At a frequency of

$$\omega = \frac{1}{R(C_{12}+C_{2G})} \, , \qquad (2\text{-}4)$$

Equation 2-2 gives a value of noise that is 1.41 times the actual value. In almost all practical cases, the frequency is much less than this, and Eq. 2-2 applies.

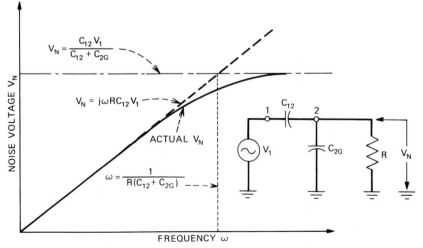

Figure 2-3. *Frequency response of capacitive coupled noise voltage.*

EFFECT OF SHIELD ON CAPACITIVE COUPLING

First consider the case where the receiver (conductor 2) has infinite resistance to ground. If a shield is placed around conductor 2, the configuration becomes that of Fig. 2-4. An equivalent circuit of the capacitive coupling between conductors is included. The voltage picked up by the shield is

$$V_S = \left(\frac{C_{1S}}{C_{1S} + C_{SG}} \right) V_1. \tag{2-5}$$

Since there is no current flow through C_{2S} the voltage picked up by conductor 2 is

$$V_N = V_S. \tag{2-6}$$

If the shield is grounded, the voltage $V_S = 0$, and the noise voltage V_N on conductor 2 is likewise reduced to zero. This case—where the center conductor does not extend beyond the shield—is an ideal situation and not typical.

In practice, the center conductor normally does extend beyond the shield, and the situation becomes that of Fig. 2-5. There, C_{12} is the capacitance between conductor 1 and the shielded conductor 2, and C_{2G} is the capacitance between conductor 2 and ground. Both of these capacitances exist because the ends of conductor 2 extend beyond the shield. Even if the shield is grounded, there is a noise voltage coupled to conductor 2. Its magnitude is expressed as follows:

$$V_N = \frac{C_{12}}{C_{12} + C_{2G} + C_{2S}} V_1. \tag{2-7}$$

The value of C_{12}, and hence V_N, in Eq. 2-7 depends on the length of conductor 2 that extends beyond the shield.

For good electric-field shielding, therefore, it is necessary to (1) *minimize the length of the center conductor that extends beyond the shield, and* (2) *provide a good ground on the shield.* A single ground connection makes a good shield ground, provided the cable is not longer than one-twentieth of a wavelength. On longer cables, multiple grounds may be necessary.

If in addition the receiving conductor has finite resistance to ground, the arrangement is that shown in Fig. 2-6. If the shield is grounded, the equivalent circuit can be simplified as shown in the figure. Any capacitance directly across the source can be neglected, since it has no effect on the noise coupling. The simplified equivalent circuit can be recognized as the same circuit analyzed in Fig. 2-1, provided C_{2G} is replaced by the sum

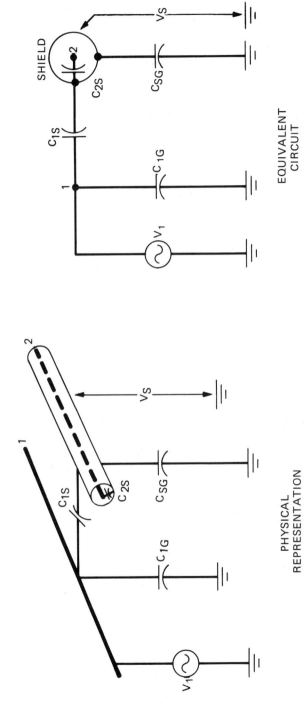

Figure 2-4. *Capacitive coupling with shield placed around receiver conductor.*

28

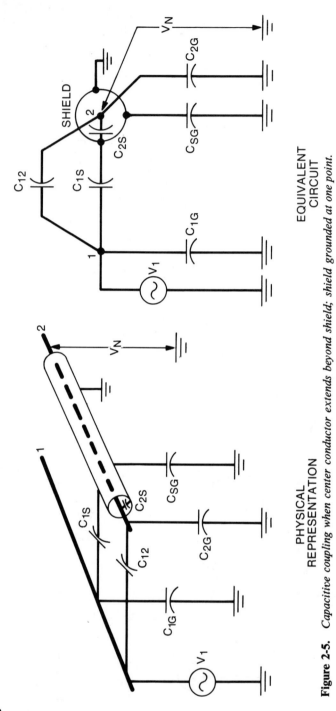

PHYSICAL REPRESENTATION

EQUIVALENT CIRCUIT

Figure 2-5. Capacitive coupling when center conductor extends beyond shield; shield grounded at one point.

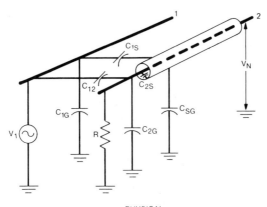

PHYSICAL
REPRESENTATION

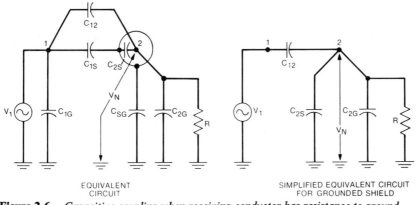

EQUIVALENT
CIRCUIT

SIMPLIFIED EQUIVALENT CIRCUIT
FOR GROUNDED SHIELD

Figure 2-6. *Capacitive coupling when receiving conductor has resistance to ground.*

of C_{2G} and C_{2S}. Therefore, if

$$R \ll \frac{1}{j\omega(C_{12} + C_{2G} + C_{2S})},$$

which is normally true, the noise voltage coupled to conductor 2 is

$$V_N = j\omega R C_{12} V_1. \tag{2-8}$$

This is the same as Eq. 2-2, which is for an unshielded cable, except that C_{12} is greatly reduced by the presence of the shield. Capacitance C_{12} now consists primarily of the capacitance between conductor 1 and the unshielded portions of conductor 2. If the shield is braided, any capacitance that exists from conductor 1 to 2 through the holes in the shield must also be included in C_{12}.

INDUCTIVE COUPLING

When a current I flows in a closed circuit, it produces a magnetic flux ϕ which is proportional to the current. The constant of proportionality is called the inductance L, hence we can write

$$\phi = LI. \tag{2-9}$$

The inductance value depends on the geometry of the circuit and the magnetic properties of the medium containing the field. Inductance has meaning only for a closed circuit. However, at times we may talk about the inductance of only a portion of a circuit. In this case, we mean the contribution that a segment of the circuit makes to the total inductance of the closed circuit.

When current flow in one circuit produces a flux in a second circuit, there is a mutual inductance M_{12} between circuits 1 and 2 defined as

$$M_{12} = \frac{\phi_{12}}{I_1}. \tag{2-10}$$

The symbol ϕ_{12} represents the flux in circuit 2 due to the current I_1 in circuit 1.

The voltage V_N induced in a closed loop of area $\bar{A}$ due to a magnetic field of flux density $\bar{B}$ can be shown (Hayt, p. 331) as

$$V_N = -\frac{d}{dt} \int_A \bar{B} \cdot \bar{A}, \tag{2-11}$$

where $\bar{B}$ and $\bar{A}$ are vectors. If the closed loop is stationary and the flux density is sinusoidally varying with time, but constant over the area of the loop, Eq. 2-11 reduces to

$$\boxed{V_N = j\omega BA \cos\theta.}^{\,*} \tag{2-12}$$

As shown in Fig. 2-7, A is the area of the closed loop, B is the rms value of

---

*Equation 2-12 is correct when the MKS system of units is being used. Flux density B is in webers per square meter (or tesla), and area A is in square meters. If B is expressed in gauss and A is in square centimeters (the CGS system), the right side of Eq. 2-12 must be multiplied by 10^{-8}.

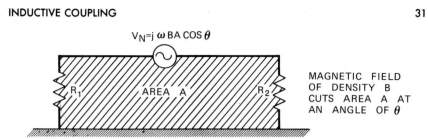

$V_N = j\,\omega\,BA\,\cos\theta$

AREA A

R_1

R_2

MAGNETIC FIELD OF DENSITY B CUTS AREA A AT AN ANGLE OF θ

Figure 2-7. *Induced noise depends on the area enclosed by the disturbed circuit.*

the sinusoidally varying flux density of frequency ω radians per second, and V_N is the rms value of the induced voltage.

This relationship can also be expressed in terms of the mutual inductance M between two circuits, as follows:

$$V_N = j\omega M I_1 = M\,\frac{di_1}{dt}. \qquad (2\text{-}13)$$

Equations 2-12 and 2-13 are the basic equations describing inductive coupling between two circuits. Figure 2-8 shows the inductive (magnetic) coupling between two circuits as described by Eq. 2-13. I_1 is the current in the interfering circuit, and M is the term that accounts for the geometry and the magnetic properties of the medium between the two circuits. The presence of ω in Eqs. 2-12 and 2-13 indicates that the coupling is directly proportional to frequency. To reduce the noise voltage, B, A, or $\cos\theta$ must

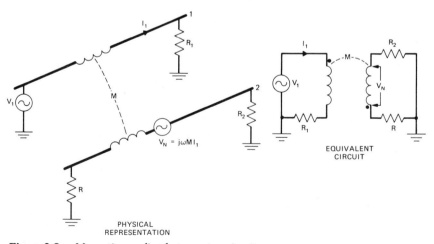

Figure 2-8. *Magnetic coupling between two circuits.*

be reduced. The B term can be reduced by physical separation of the circuits or by twisting the source wires, provided the current flows in the twisted pair and not through the ground plane. The conditions necessary for this are covered in a later section. Under these conditions, twisting causes the B fields from each of the wires to cancel. The area of the receiver circuit can be reduced by placing the conductor closer to the ground plane (if the return current is through the ground plane) or by using two conductors twisted together (if the return current is on one of the pair instead of the ground plane). The $\cos\theta$ term can be reduced by proper orientation of the source and receiver circuits.

If may be helpful to note some differences between magnetic and electric field coupling. First, reducing the impedance of the receiver circuit in a magnetically coupled situation does not decrease the pickup as it does in the case of electric field coupling. Second, in the case of magnetic field coupling, the noise voltage is produced in series with the receiver conductors, whereas in the case of electric field coupling, the noise voltage is produced between the receiver conductor and ground.

If an ungrounded and nonmagnetic shield is now placed around conductor 2, the circuit becomes that of Fig. 2-9, where M_{1S} is the mutual inductance between conductor 1 and the shield. Since the shield has no effect on the geometry or magnetic properties of the medium between circuits 1 and 2 it has no effect on the voltage induced into conductor 2. The shield does, however, pick up a voltage due to the current in conductor 1.

$$V_S = j\omega M_{1S} I_1. \qquad (2\text{-}14)$$

A ground connection on one end of the shield does not change the situation. *It can, therefore, be concluded that a shield placed around a conductor and grounded at one end has no effect on the magnetically induced voltage in that conductor.*

MAGNETIC COUPLING BETWEEN SHIELD AND INNER CONDUCTOR

Before continuing the discussion of inductive coupling, it will be necessary to calculate the magnetic coupling between a hollow conducting tube and any conductors placed inside the tube. This concept is fundamental to a discussion of inductive shielding and will be needed later.

First consider the magnetic field produced by a tubular conductor carrying a uniform axial current, as shown in Fig. 2-10. If the hole in the tube is concentric with the outside of the tube, there is no magnetic field in the cavity and the total magnetic field is external to the tube (Smythe, p. 278). Now let a conductor be placed inside the tube to form a coaxial

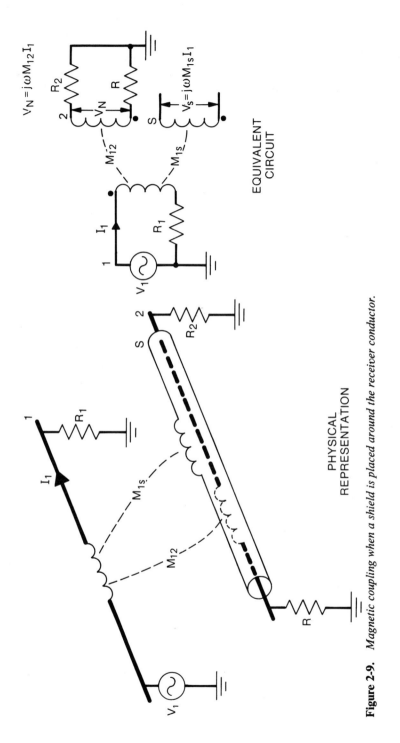

Figure 2-9. *Magnetic coupling when a shield is placed around the receiver conductor.*

$V_N = j\omega M_{12}I_1$

$V_s = j\omega M_{1s}I_1$

EQUIVALENT CIRCUIT

PHYSICAL REPRESENTATION

33

cable, as shown in Fig. 2-11. All of the flux ϕ due to the current I_S in the shield tube encircles the inner conductor. The inductance of the shield is equal to

$$L_S = \frac{\phi}{I_S}. \tag{2-15}$$

The mutual inductance between the shield and the inner conductor is

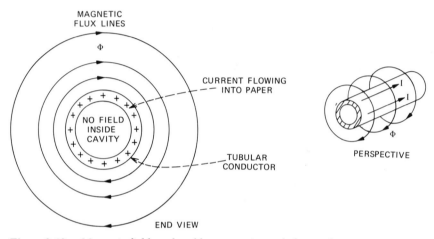

Figure 2-10. *Magnetic field produced by current in a tubular conductor.*

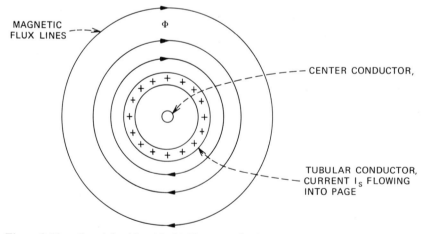

Figure 2-11. *Coaxial cable with shield current flowing.*

equal to

$$M = \frac{\phi}{I_S}.$$ (2-16)

Since all the flux produced by the shield current encircles the center conductor, the flux in these two equations is the same. The mutual inductance between the shield and center conductor is therefore equal to the self inductance of the shield

$$\boxed{M = L_S.}$$ (2-17)

Equation 2-17 is a most important result and one that we will often have occasion to refer to. It was derived to show that the mutual inductance between the shield and the center conductor is equal to the shield inductance. Based on the reciprocity of mutual inductance (Hayt, p. 321), the inverse must also be true. That is, the mutual inductance between the center conductor and the shield is equal to the shield inductance.

The validity of Eq. 2-17 depends only on the fact that there is no magnetic field in the cavity of the tube due to shield current. The requirements for this to be true are that the tube be cylindrical and the current density be uniform around the circumference of the tube. Equation 2-17 applies regardless of the position of the center conductor within the tube. In other words, the two conductors do not have to be coaxial.

The voltage V_N induced into the center conductor due to a current I_S in the shield can now be calculated. Assume that the shield current is produced by a voltage V_S induced into the shield from some other circuit. Figure 2-12 shows the circuit being considered; L_S and R_S are the inductance and resistance of the shield. The voltage V_N is equal to

$$V_N = j\omega M I_S.$$ (2-18)

The current I_S is equal to

$$I_S = \frac{V_S}{L_S}\left(\frac{1}{j\omega + R_S/L_S}\right).$$ (2-19)

Therefore,

$$V_N = \left(\frac{j\omega M V_S}{L_S}\right)\left(\frac{1}{j\omega + R_S/L_S}\right).$$ (2-20)

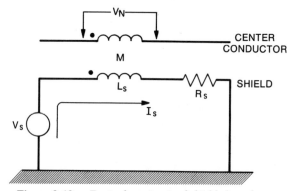

Figure 2-12. *Equivalent circuit of shielded conductor.*

Since $L_S = M$ (from Eq. 2-17),

$$V_N = \left(\frac{j\omega}{j\omega + R_S/L_S} \right) V_S. \qquad (2\text{-}21)$$

A plot of Eq. 2-21 is shown in Fig. 2-13. The break frequency for this curve is defined as the shield cutoff frequency (ω_c) and occurs at

$$\omega_c = \frac{R_S}{L_S}, \qquad \text{or} \qquad f_c = \frac{R_S}{2\pi L_S}. \qquad (2\text{-}22)$$

The noise voltage induced into the center conductor is zero at dc and increases to almost V_S at a frequency of $5R_S/L_S$ rad/s. Therefore, if shield current is allowed to flow, a voltage is induced into the center conductor that nearly equals the shield voltage at frequencies greater than five times the shield cutoff frequency.

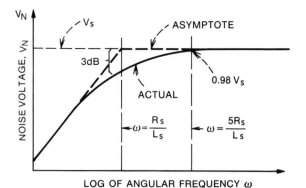

Figure 2-13. *Noise voltage in center conductor of coaxial cable due to shield current.*

This is a very important property of a conductor inside a shield. Measured values of the shield cutoff frequency and five times this frequency are tabulated in Table 2-1 for various cables. For most cables, five times the shield cutoff frequency is near the high end of the audio-frequency band. The aluminum-foil-shielded cable listed has a much higher shield cutoff frequency than any other. This is due to the increased resistance of its thin aluminum-foil shield.

Table 2-1   Measured Values of Shield Cutoff Frequency (f_c)

Cable	Impedance (Ω)	Cutoff frequency (kHz)	Five times cutoff frequency (kHz)	Remarks
Coaxial cable				
RG-6A	75	0.6	3.0	Double shielded
RG-213	50	0.7	3.5	
RG-214	50	0.7	3.5	Double shielded
RG-62A	93	1.5	7.5	
RG-59C	75	1.6	8.0	
RG-58C	50	2.0	10.0	
Shielded twisted pair				
754E	125	0.8	4.0	Double shielded
24 Ga.	—	2.2	11.0	
22 Ga.[a]	—	7.0	35.0	Aluminum-foil shield
Shielded single				
24 Ga.	—	4.0	20.0	

[a]One pair out of an 11 pair cable (Belden 8775).

SHIELDING TO PREVENT MAGNETIC RADIATION

To prevent radiation, the source of the interference may be shielded. Fig. 2-14 shows the electric and magnetic fields surrounding a current-carrying

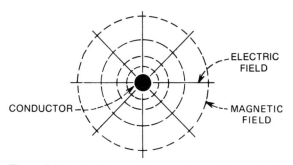

Figure 2-14.   *Fields around a current carrying conductor.*

conductor located in free space. If a shield grounded at one point is placed around the conductor, the electric field lines will terminate on the shield but there will be very little effect on the magnetic field. This is shown in Fig. 2-15. If a shield current equal and opposite to that in the center conductor is made to flow on the shield, it generates an equal but opposite external magnetic field. This field cancels the magnetic field caused by the center conductor external to the shield. This results in the condition shown in Fig. 2-16, with no fields external to the shield.

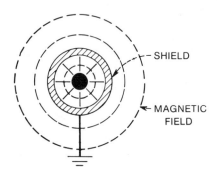

SHIELD

MAGNETIC
FIELD

Figure 2-15. *Fields around shielded conductor; shield grounded at one point.*

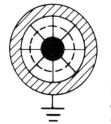

Figure 2-16. *Fields around shielded conductor; shield grounded and carrying a current equal to the conductor current, but in the opposite direction.*

Figure 2-17 shows a circuit which is grounded at both ends and carries a current I_1. To prevent magnetic field radiation from this circuit the shield must be grounded at both ends and the return current must flow from A to B in the shield (I_S in the figure) instead of in the ground plane (I_G in the figure). But why should the current return from point A to B through the shield instead of through the zero-resistance ground plane? The equivalent circuit can be used to analyze this configuration. By writing a mesh equation around the ground loop ($A - R_S - L_S - B - A$) the shield current I_S can be determined:

$$0 = I_S \left(j\omega L_S + R_S \right) - I_1 (j\omega M), \qquad (2\text{-}23)$$

where M is the mutual inductance between the shield and center conductor and as previously shown (Eq. 2-17), $M = L_S$. Making this substitution and rearranging produces this expression for I_S:

$$I_S = I_1\left(\frac{j\omega}{j\omega + R_S/L_S}\right) = I_1\left(\frac{j\omega}{j\omega + \omega_c}\right). \qquad (2\text{-}24)$$

As can be seen from the above equation, if the frequency is much above the shield cutoff frequency ω_c the shield current approaches the center conductor current. Because of the mutual inductance between the shield and center conductor, therefore, the shield provides a return path with lower total circuit inductance than the ground plane at high frequency. As the frequency decreases below $5\omega_c$, the cable provides less and less magnetic shielding since more of the current returns via the ground plane.

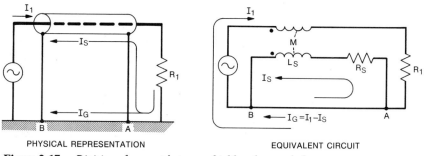

PHYSICAL REPRESENTATION   EQUIVALENT CIRCUIT

Figure 2-17. *Division of current between shield and ground plane.*

To prevent radiation of a magnetic field from a conductor grounded at both ends, it should be shielded and the shield should be grounded at both ends. This provides good magnetic field shielding at frequencies considerably above the shield cutoff frequency. This reduction in the radiated magnetic field is not because of the magnetic shielding properties of the shield as such. Rather, the return current on the shield generates a field that cancels the conductor's field.

If the ground is removed from one end of the circuit, as shown in Fig. 2-18, then the shield should not be grounded at that end since the return current must now all flow on the shield. This is true especially at frequencies less than the shield cutoff frequency. Grounding both ends of the shield, in this case, reduces the shielding since some current would return via the ground plane.

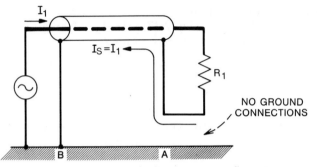

Figure 2-18. *Without ground at far end, all return current flows through shield.*

SHIELDING A RECEIVER AGAINST MAGNETIC FIELDS

The best way to protect against magnetic fields at the receiver is to decrease the area of the receiver loop. The area of interest is the total area enclosed by current flow in the receiver circuit. An important consideration is the path taken by the current in returning to the source. Quite often the current returns by a path other than the one intended by the designer, and therefore, the area of the loop changes. If a nonmagnetic shield placed around a conductor causes the current to return over a path that encloses a smaller area, then some protection against magnetic fields will have been provided by the shield. This protection, however, is due to the reduced loop area and not to any magnetic shielding properties of the shield.

Figure 2-19 illustrates the effect of a shield on the loop area of a circuit. In Fig. 2-19A, the source V_s is connected to the load R_L by a single conductor, using a ground return path. The area enclosed by the current is the rectangle between the conductor and the ground plane. In Fig. 2-19B, a shield is placed around the conductor and grounded at both ends. If the current returns through the shield rather than the ground plane, the area of the loop is decreased and a degree of magnetic protection is provided. The current returns through the shield if the frequency is greater than five times the shield cutoff frequency as previously shown. A shield placed around the conductor and grounded at one end only as shown in Fig. 2-19C does not change the loop area, and therefore, provides no magnetic protection.

The arrangement of Fig. 2-19B does not protect against magnetic fields at frequencies below the shield cutoff frequency since then most of the current returns through the ground plane and not through the shield. This circuit should be avoided at low frequencies for two other reasons: (1) since the shield is one of the circuit conductors, any noise current in it will produce an *IR* drop in the shield and appear to the circuit as a noise voltage, and (2) if there is a difference in ground potential between the two

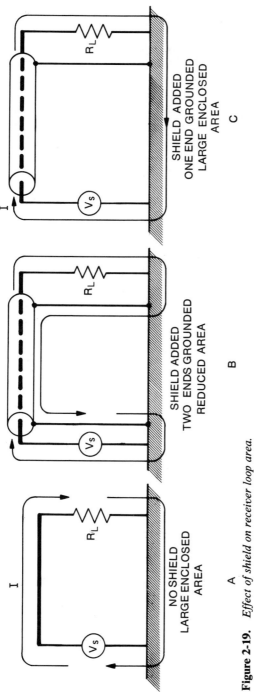

Figure 2-19. *Effect of shield on receiver loop area.*

ends of the shield, this too will show up as a noise voltage in the circuit.*

Whenever a circuit is grounded at both ends, only a limited amount of magnetic field protection is possible because of the large noise current induced in the ground loop. Since this current flows through the signal conductor, a noise voltage is produced in the shield, equal to the shield current times the shield resistance. This is shown in Fig. 2-20. The current I_S is the noise current in the shield due to a ground differential or to external noise coupling. If voltages are added around the input loop, the following expression is obtained:

$$V_{IN} = -j\omega M I_S + j\omega L_S I_S + R_S I_S. \qquad (2\text{-}25)$$

Since $L_S = M$, as was previously shown

$$V_{IN} = R_S I_S. \qquad (2\text{-}26)$$

Whenever shield current flows a noise voltage is produced in the shield due to the $I_S R_S$ voltage drop.

Even if the shield is grounded at only one end, shield noise currents may

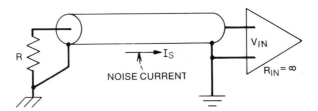

PHYSICAL REPRESENTATION

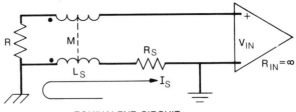

EQUIVALENT CIRCUIT

Figure 2-20. *Effect of noise current flowing in the shield of a coaxial cable.*

*See Chapter 3, p. 79, for further discussion of a shielded cable grounded at both ends.

still flow due to capacitive coupling to the shield. *Therefore, for maximum noise protection at low frequencies, the shield should not be one of the signal conductors, and one end of the circuit must be isolated from ground.*

EXPERIMENTAL DATA

The magnetic field shielding properties of various cable configurations were measured and compared. The test setup is shown in Fig. 2-21, and the test results are tabulated in Figs. 2-22 and 2-23. The frequency (50 kHz) is greater than five times the shield cutoff frequency for all the cables tested. The cables shown in Figs. 2-22 and 2-23 represent tests cables shown as *L2* in Fig. 2-21.

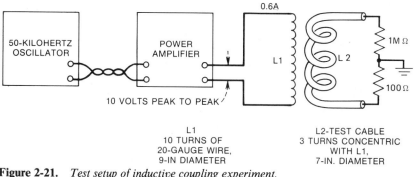

Figure 2-21. *Test setup of inductive coupling experiment.*

In circuits *A* through *F* (Fig. 2-22), both ends of the circuit are grounded. They provide much less magnetic field attenuation than do circuits *G* through *K* (Fig. 2-23), where only one end is grounded.

Circuit *A* in Fig. 2-22 provides essentially no magnetic field shielding. The actual noise voltage measured across the one megohm resistor in this case was 0.8 V. The pickup in configuration *A* is used as a reference, and is called 0 dB, to compare the performance of all the other circuits. In circuit *B*, the shield is grounded at one end; this has no effect on the magnetic shielding. Grounding the shield at both ends as in configuration *C* provides some magnetic field protection because the frequency is above the shield cutoff frequency. The protection would be even greater if it were not for the ground loop formed by grounding both ends of the circuit. The magnetic field induces a large noise current into the low impedance ground loop consisting of the cable shield and the two ground points. The shield noise current then produces a noise voltage in the shield, as was shown in the preceding section.

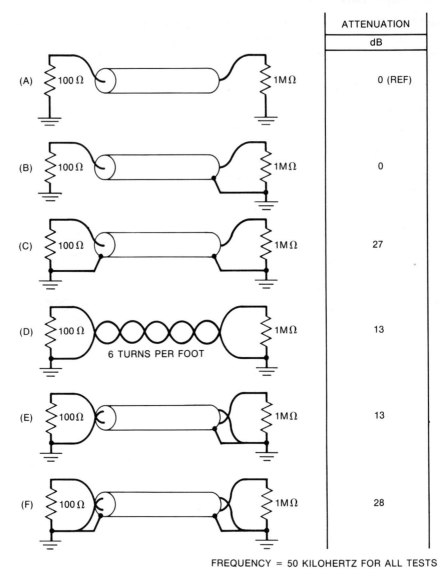

| | ATTENUATION |
	dB
(A) 100 Ω 1MΩ	0 (REF)
(B) 100 Ω 1MΩ	0
(C) 100 Ω 1MΩ	27
(D) 100 Ω 1MΩ 6 TURNS PER FOOT	13
(E) 100Ω 1MΩ	13
(F) 100 Ω 1MΩ	28

FREQUENCY = 50 KILOHERTZ FOR ALL TESTS

Figure 2-22. *Results of inductive coupling experiment; all circuits grounded at both ends.*

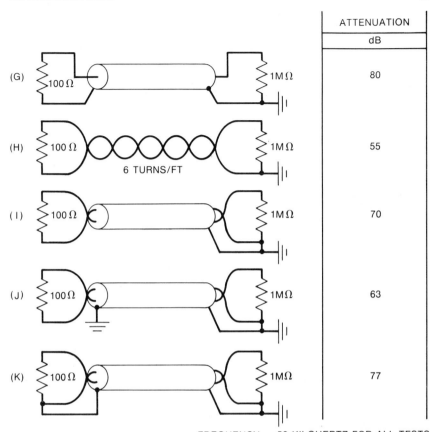

FREQUENCY = 50 KILOHERTZ FOR ALL TESTS

Figure 2-23. *Results of inductive coupling experiment; all circuits are grounded at one end only.*

Use of a twisted pair as in circuit *D* should provide much greater magnetic field noise reduction, but its effect is defeated by the ground loop formed by circuit grounds at both ends. This can clearly be seen by comparing the attenuation of circuit *H* to that of circuit *D*. Adding a shield with one end grounded, to the twisted pair as in *E* has no effect. Grounding the shield at both ends as in *F* provides additional protection, since the low-impedance shield shunts some of the magnetically induced ground-loop current away from the signal conductors. In general, however, none of the circuit configurations in Fig. 2-22 provide good magnetic field protection because of the ground loops. If the circuit must be grounded at both ends, configurations *C* or *F* should be used.

Circuit G shows a significant improvement in magnetic field shielding. This is due to the very small loop area formed by the coaxial cable and the fact that there is no ground loop to defeat the shielding. The coax provides a very small loop area since the shield can be represented by an equivalent conductor located on its center axis. This effectively locates the shield at or very near the axis of the center conductor.

It was expected that the twisted pair of circuit H would provide considerably more shielding than the 55 dB shown. The reduced shielding is due to the fact that some electric-field coupling is now beginning to show up. This can be seen in circuit I, where attenuation increases to 70 dB by placing a shield around the twisted pair. The fact that attenuation in circuit G is better than in I indicates that in this case the particular coaxial cable presents a smaller loop area to the magnetic field than does the twisted pair. This, however, is not necessarily true in general. Increasing the number of turns per foot for either of the twisted pairs (H or I) would reduce the pickup. In general, circuit I is preferred over circuit G for low-frequency magnetic shielding since in I the shield is not also one of the signal conductors.

Grounding both ends of the shield as in circuit J decreases the shielding slightly. This is probably due to the high shield current in the ground loop formed by the shield inducing unequal voltages in the two center conductors. Circuit K provides more shielding than I since it combines the features of the coax G with those of the twisted pair I. Circuit K is not normally desirable since any noise voltages or currents that do get on the shield can flow down the signal conductor. It is almost always better to connect the shield and signal conductors together at just one point. That point should be such that noise current from the shield does not have to flow down the signal conductor to get to ground.

SHIELD FACTOR

The amount of shielding between two circuits can be expressed in terms of a shield factor. The shield factor (η) is defined as the ratio of induced voltage in the disturbed circuit (receiver), after the shield is introduced, to the same induced voltage without the shield:

$$\eta = \frac{V_N \text{ (shield in place)}}{V_N \text{ (no shield)}} . \tag{2-27}$$

A generalized shielding arrangement is depicted in Fig. 2-24. It consists of: (1) a disturbing conductor, (2) a shield conductor, and (3) a disturbed conductor. For the arrangement illustrated, it can be shown that:

$$\eta = 1 - \frac{Z_{12}Z_{23}}{Z_{13}Z_{22}} \,. \tag{2-28}$$

Impedance Z_{22} is the self impedance of circuit 2, and Z_{12}, Z_{13} and Z_{23} are the mutual impedances between circuits 1 and 2, 1 and 3, and 2 and 3, respectively.

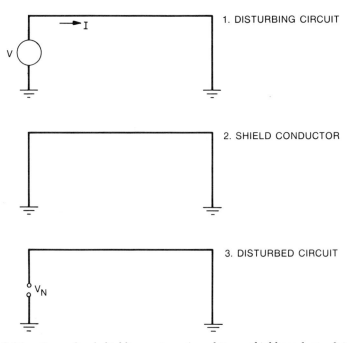

Figure 2-24. *Generalized shielding system. An arbitrary shield conductor 2 is placed between the noise source 1 and the receiver conductor 3.*

EXAMPLE OF SELECTIVE SHIELDING

The shielded loop antenna is an example where the electric field is selectively shielded while the magnetic field is unaffected. Such an antenna is useful in radio direction finders. It can also decrease the antenna noise pickup in broadcast receivers. The latter effect is significant because the majority of local noise sources generate a predominantly electric field. Figure 2-25A shows the basic loop antenna. From Eq. 2-12, the magnitude of the voltage produced in the loop by the magnetic field is

$$V_m = 2\pi fBA \cos\theta. \tag{2-29}$$

The angle θ is measured between the magnetic field and a perpendicular to the plane of the loop. The loop, however, also acts as a vertical antenna and picks up a voltage due to an incident electric field. This voltage is equal to the E field times the effective height of the antenna. For a circular single-loop antenna, the effective height is $2\pi A/\lambda$ (ITT, p. 25-6). The induced voltage due to the electric field becomes

$$V_e = \frac{2\pi AE}{\lambda} \cos\theta'. \tag{2-30}$$

The angle θ' is measured between the electric field and the plane of the loop.

To eliminate pickup from the electric field, the loop could be shielded as shown in Fig. 2-25B. However, this configuration allows shield current to flow, which will cancel the magnetic field as well as the electric field. To preserve the magnetic sensitivity of the loop, the shield must be broken to prevent the flow of shield current. This can be done as shown in Fig. 2-25C by breaking the shield at the top. The resulting antenna responds only to the magnetic field component of an applied wave.

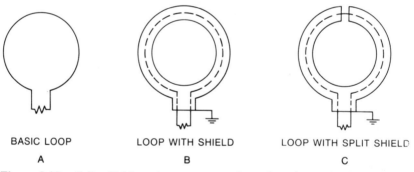

BASIC LOOP   LOOP WITH SHIELD   LOOP WITH SPLIT SHIELD

A   B   C

Figure 2-25. *Split shield on loop antenna selectively reduces electric field while passing magnetic field.*

COAXIAL CABLE VERSUS SHIELDED TWISTED PAIR

When comparing coaxial cable with a shielded twisted pair, it is important to recognize the usefulness of both types of cable from a propagation point of view, irrespective of their shielding characteristics. This is shown in Fig.

2-26. Shielded twisted pairs are very useful at frequencies below 100 kHz. In some applications, the frequency may reach as high as 10 MHz. Above 1 MHz, the losses in the shielded twisted pair increase considerably.

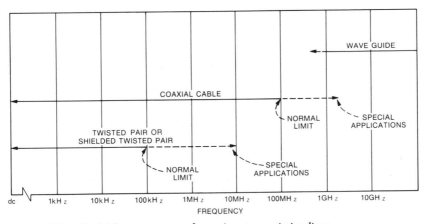

Figure 2-26. *Useful frequency range for various transmission lines.*

On the other hand coaxial cable has a more uniform characteristic impedance with lower losses. It is useful, therefore, from zero frequency (dc) up to VHF frequencies, with some applications extending up to UHF. Above a few hundred megahertz, the losses in coaxial cable become large, and waveguide becomes more practical. A shielded twisted pair has more capacitance than a coaxial cable and, therefore, is not as useful at high frequencies or in high-impedance circuits.

A coaxial cable grounded at one point provides a good degree of protection from capacitive pickup. But if a noise current flows in the shield, a noise voltage is produced. Its magnitude is equal to the shield current times the shield resistance. Since the shield is part of the signal path, this noise voltage appears as noise in series with the input signal. A double-shielded, or triaxial, cable with insulation between the two shields can eliminate the noise produced by the shield resistance. The noise current flows in the outer shield, and the signal current flows in the inner shield. The two currents (signal and noise), therefore, do not flow through a common impedance.

Unfortunately, triaxial cables are expensive and awkward to use. A coaxial cable at high frequencies, however, acts as a triaxial cable due to skin effect. For a typical shielded cable, skin effect becomes important at about 1 MHz. The noise current flows on the outside surface of the shield while the signal current flows on the inside surface. For this reason a coaxial cable is better for use at high frequencies.

A shielded twisted pair has characteristics similar to a triaxial cable and is not as expensive or awkward. The signal current flows in the two inner conductors, and any noise currents flow in the shield. Common-resistance coupling is eliminated. In addition, any shield current is coupled equally into both inner conductors by mutual inductance, and the voltages therefore cancel.

An unshielded twisted pair, unless it is balanced, provides very little protection against capacitive pickup, but it is very good for protection against magnetic pickup. The shielded twisted pair provides the best shielding for low-frequency signals, in which magnetic pickup is the major problem. The effectiveness of twisting increases as the number of twists per unit length increases.

BRAIDED SHIELDS

Most cables are actually shielded with braid rather than with a solid conductor. The advantages of braid are flexibility, durability, strength, and long flex life. Braids, however, typically provide only 60–90% coverage and are less effective as shields than solid conductors. Braided shields usually provide just slightly reduced electric field shielding (except at UHF frequencies) but greatly reduced magnetic field shielding. The reason is that braid distorts the uniformity of the shield current. A braid is typically from 5 to 30 dB less effective than a solid shield for protecting against magnetic fields.

At higher frequencies, the effectiveness of the braid decreases further. This is because the braid holes become larger compared to a wavelength, as the frequency increases. Multiple shields offer more protection, but with higher cost and less flexibility. Cables with double or even triple shields are used in some critical applications. Recently, cables with solid aluminum foil shields have become available. These shields provide almost 100% coverage and more effective shielding. They are not as strong as braid, however, and usually have a higher shield cutoff frequency due to their higher shield resistance.

UNIFORMITY OF SHIELD CURRENT

The magnetic shielding previously discussed depends on a uniform distribution of the longitudinal shield current around the shield circumference. Solid shields such as aluminum foil produce the most uniform shield current distribution, and therefore, provide the best magnetic shielding if the frequency is above the shield cutoff frequency. Braided shields are considerably less effective for magnetic shielding since their current distribution is less uniform than that of a solid shield. The braid can be plated—typically with solder or silver—and current flow is more uniform due to better conductor-to-conductor contact. Unplated shields tend to build up an oxide coating and have poor electrical contact between individual braid wires.

The magnetic shielding effectiveness near the ends of the cable depends on the way the braid is terminated. A pigtail connection, Fig. 2-27, causes the shield current to be concentrated on one side of the shield. For maximum protection, the shield should be terminated uniformly around its cross section. This can be accomplished by using a coaxial connector such as the BNC, UHF, or Type N connectors. Such a connector, shown in Fig. 2-28, provides 360° electrical contact to the shield. A coaxial termination also provides complete coverage of the inner conductor, preserving the integrity of electric field shielding.

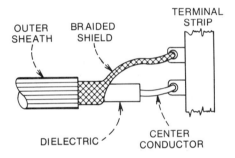

Figure 2-27. *Pigtail shield connection concentrates current on one side of shield.*

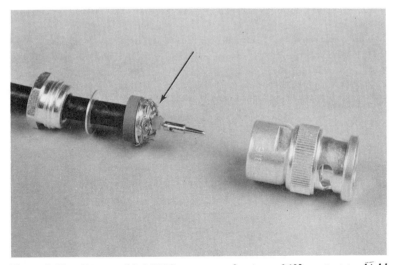

Figure 2-28. *Disassembled BNC connector showing a 360° contact to shield.*

SUMMARY

- Electric fields are much easier to guard against than magnetic fields.
- The use of nonmagnetic shields around conductors provides no magnetic shielding per se.
- A shield grounded at one or more points shields against electric fields.
- The key to magnetic shielding is to decrease the area of the loop. To do that use twisted pair, or use coaxial cable if the current return is through the shield instead of in the ground plane.
- For a coaxial cable grounded at both ends, virtually all of the return current flows in the shield at frequencies above five times the shield cutoff frequency.
- To prevent radiation from a conductor, a shield grounded at both ends is useful above the shield cutoff frequency.
- Only a limited amount of magnetic shielding is possible in a receiver circuit that is grounded at both ends, due to the ground loop formed.
- Any shield in which noise currents flow should not be part of the signal path. Use shielded twisted pair or triaxial cable at low frequencies.
- At high frequencies, a coaxial cable acts as a triaxial cable due to skin effect.
- The shielding effectiveness of twisted pair increases as the number of twists per unit length increase.
- The magnetic shielding effects listed above require a cylindrical shield with uniform distribution of shield current over the circumference of the shield.

BIBLIOGRAPHY

Belden Engineering Staff, *Electronics Cable Handbook*, Howard W. Sams & Co., New York, 1966.

Bell Laboratories, *Physical Design of Electronic Systems*, Vol. 1, Chapter 10 (Electrical Interference), Prentice-Hall, Englewood Cliffs, N.J., 1970.

Buchman, A. S., "Noise Control in Low Level Data Systems," *Electromechanical Design*, September, 1962.

Ficchi, R. O., *Electrical Interference*, Hayden Book Co., New York, 1964.

Ficchi, R. O., *Practical Design For Electromagnetic Compatibility*, Hayden Book Co., New York, 1971.

Frederick Research Corp., *Handbook on Radio Frequency Interference*, Vol. 3, (Methods of Electromangetic Interference Suppression), Frederick Research Corp., Wheaton, Maryland, 1962.

Hayt, W. H., Jr., *Engineering Electromagnetics*, Third Edition; McGraw-Hill, New York, 1974.

ITT, *Reference Data for Radio Engineers*, Fifth Edition; Howard W. Sams & Co., New York, 1968.

Morrison, R., *Grounding and Shielding Techniques in Instrumentation*; Wiley, New York, 1967.

Nalle, D., "Elimination of Noise in Low Level Circuits," *ISA Journal*, Vol. 12, August, 1965.

Smythe, W. R., *Static and Dynamic Electricity*, McGraw-Hill, New York, 1924.

Timmons, F., "Wire or Cable Has Many Faces, Part 2," *EDN*, March, 1970.

Trompeter, E., "Cleaning Up Signals With Coax," *Electronic Products Magazine*, July 16, 1973.

White, D. R. J., *Electromagnetic Interference and Compatibility*, Vol. 3, (EMI Control Methods and Techniques), Don White Consultants Inc., Germantown, Maryland, 1973.

3 GROUNDING

Grounding is one of the primary ways to minimize unwanted noise and pickup. Proper use of grounding and shielding, in combination, can solve a large percentage of all noise problems. A good grounding system must be designed just like the rest of the circuit. It is sometimes difficult to convince one's self that expensive engineering time should be spent on the minute details of where every circuit should be grounded. However in the long run, time and money are usually saved by not having to solve mysterious interference problems that turn up when the equipment is later built and tested.

The grounding principles covered here are just as applicable to large complex electronic systems as they are to individual circuits on a single printed wiring board. There are two basic objectives involved in designing good grounding systems. The first is to *minimize the noise voltage generated by currents from two or more circuits flowing through a common ground impedance.* The second is to *avoid creating ground loops which are susceptible to magnetic fields and differences in ground potential.* Grounding, if done improperly however, can become a primary means of noise coupling.

In the most general sense a ground can be defined as an equipotential* point or plane which serves as a reference voltage for a circuit or system. It may or may not be at earth potential. If the ground is connected to the earth through a low impedance path, it can then be called an earth ground. There are two common reasons for grounding a circuit: (1) for safety, and (2) to provide an equipotential reference for signal voltages. Safety grounds are always at earth potential, whereas signal grounds are usually but not necessarily at earth potential. In many cases, a safety ground is required at a point which is unsuitable for a signal ground, and this may complicate the noise problem.

SAFETY GROUNDS

Safety considerations require the chassis or enclosure for electric equipment to be grounded. Why this is so can be seen in Fig. 3-1. In the left-hand diagram Z_1 is the stray impedance between a point at potential

---

*A point where the voltage does not change regardless of the amount of current supplied to it or drawn from it.

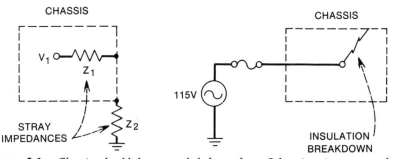

Figure 3-1. *Chassis should be grounded for safety. Otherwise, it may reach a dangerous voltage level through stray impedances (left) or insulation breakdown (right).*

V_1 and the chassis, and Z_2 is the stray impedance between the chassis and ground. The potential of the chassis is determined by impedances Z_1 and Z_2 acting as a voltage divider. The chassis potential is

$$V_{chassis} = \left(\frac{Z_2}{Z_1 + Z_2} \right) V_1. \tag{3-1}$$

The chassis could be a relatively high potential and be a shock hazard, since its potential is determined by the relative values of the stray impedances over which there is very little control. If the chassis is grounded, however, its potential is zero since Z_2 becomes zero.

The right-hand diagram of Fig. 3-1 shows a second and far more dangerous situation: a fused ac line entering an enclosure. If there should be an insulation breakdown such that the ac line comes in contact with the chassis, the chassis would then be capable of delivering the full current capacity of the fused circuit. Anyone coming in contact with the chassis and ground would be connected directly across the ac power line. If the chassis is grounded, however, such an insulation breakdown will draw a large current from the ac line and cause the fuse to blow, thus removing the voltage from the chassis.

In the United States, ac power distribution and wiring standards are contained in the National Electrical Code. One requirement of this code specifies that 115-V ac power distribution in homes and buildings must be a three wire system, as shown in Fig. 3-2. Load current flows through the hot (black) wire, which is fused, and returns through the neutral (white) wire. In addition, a safety ground (green) wire must be connected to all equipment enclosures and hardware. The only time the green wire carries current is during a fault, and then only momentarily until the fuse or breaker opens the circuit. Since no load current flows in the safety ground, it has no IR drop and the enclosures connected to it are always at ground

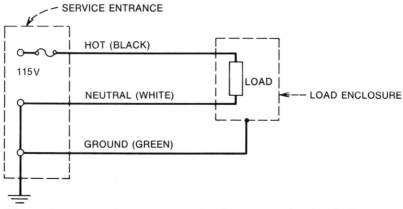

Figure 3-2. *Standard 115-V ac power distribution circuit has three leads.*

potential. The National Electrical Code specifies that the neutral and safety ground shall be connected together at only one point, and this point shall be at the main service entrance. To do otherwise would allow some of the neutral current to return on the ground conductor. A combination 115/230-V system is similar, except an additional hot wire (red) is added, as shown in Fig. 3-3. If the load requires only 230 V the neutral (white) wire shown in Fig. 3-3 is not required.

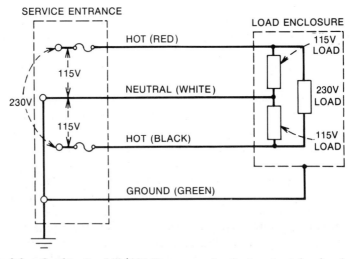

Figure 3-3. *Combination 115/230-V ac power distribution circuit has four leads.*

SIGNAL GROUNDS

Signal grounds generally fall into one of two classes: (1) single point grounds, and (2) multipoint grounds. These schemes are shown in Figs. 3-4 and 3-5. There are two subclasses of single point grounds—those with series connections and those with parallel connections. The series connection is also called a common ground system, and the parallel connection is called a separate ground system.

In the following discussion of grounding techniques, two key points should be kept in mind:

1. All conductors have a finite impedance, generally consisting of both resistance and inductance. At 11 kHz, a straight length of 22-gauge wire one inch above a ground plane has more inductive reactance than resistance.

2. Two physically separated ground points are seldom at the same potential.

The ac power ground is of little practical value as a signal ground. The voltage measured between two points on the power ground is typically hundreds of millivolts, and in some cases many volts. This is excessive for low-level signal circuits. A single-point connection to the power ground is usually required for safety, however.

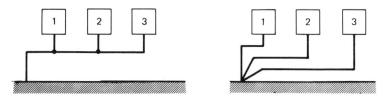

SERIES CONNECTION   PARALLEL CONNECTION

Figure 3-4. *Two types of single point grounding connections.*

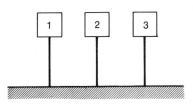

Figure 3-5. *Multipoint grounding connections.*

SINGLE POINT GROUND SYSTEMS

From a noise point of view, the most undesirable ground system is the common ground system shown in Fig. 3-6. This is a series connection of all the individual circuit grounds. The resistances shown represent the impedance of the ground conductors and I_1, I_2, and I_3 are the ground currents of circuits 1, 2, and 3, respectively. Point A is not at zero potential but is at a potential of

$$V_A = (I_1 + I_2 + I_3)R_1, \tag{3-2}$$

and point C is at a potential of

$$V_C = (I_1 + I_2 + I_3)R_1 + (I_2 + I_3)R_2 + I_3R_3. \tag{3-3}$$

Although this circuit is the least desirable grounding system, it is probably the most widely used because of its simplicity. For noncritical circuits it may be perfectly satisfactory. This system should not be used between circuits operating at widely different power levels, since the high-level stages produce large ground currents which in turn adversely affect the low-level stage. When this system is used, the most critical stage should be the one nearest the primary ground point. Note that point A in Fig. 3-6 is at a lower potential than point B or C.

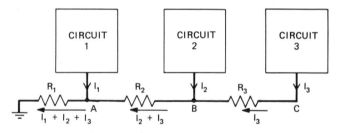

Figure 3-6. *Common ground system is a series ground connection and is undesirable from a noise standpoint but has the advantage of simple wiring.*

The separate ground system (parallel connection) shown in Fig. 3-7 is the most desirable at low frequencies. That is because there is no cross coupling between ground currents from different circuits. The potentials at points A and C, for example, are as follows:

$$V_A = I_1R_1, \tag{3-4}$$

$$V_C = I_3R_3. \tag{3-5}$$

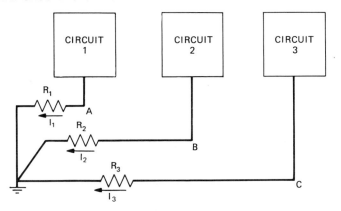

Figure 3-7. *Separate ground system is a parallel ground connection and provides good low-frequency grounding, but is mechanically cumbersome.*

The ground potential of a circuit is now a function of the ground current and impedance of that circuit only. This system is mechanically cumbersome, however, since in a large system an unreasonable amount of wire is necessary.

A second limitation of the separate ground system occurs at high frequencies, where the inductances of the ground conductors increase the ground impedance and also produce inductive coupling between the ground leads. Parasitic capacitance between the ground conductors also allows coupling between the grounds. At still higher frequencies the impedance of the ground wires can be very high if the length coincides with odd multiples of a quarter wavelength. Not only will these grounds have large impedance, but they will also act as antennas and radiate noise. Ground leads should always be kept shorter than one-twentieth of a wavelength to prevent radiation and to maintain a low impedance.

MULTIPOINT GROUND SYSTEMS

The multipoint ground system is used at high frequencies to minimize the ground impedance. In this system, shown in Fig. 3-8, circuits are connected to the nearest available low-impedance ground plane, usually the chassis. The low ground impedance is due primarily to the lower inductance of the ground plane. The connections between each circuit and the ground plane should be kept as short as possible to minimize their impedance. In very high frequency circuits the length of these ground leads must be kept to a small fraction of an inch. Multipoint grounds should be avoided at low frequencies since ground currents from all circuits flow through a common ground impedance—the ground plane. At high

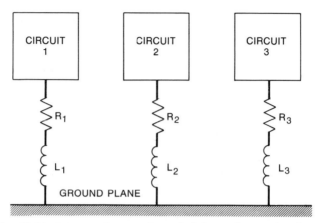

Figure 3-8. *Multipoint ground system is good choice at frequencies above 10 MHz. Impedances R_1–R_3 and L_1–L_3 should be minimized.*

frequencies, the common impedance of the ground plane can be reduced by silver plating the surface. Increasing the thickness of the ground plane has no effect on its high frequency impedance, since current flows only on the surface due to skin effect.

Normally at frequencies below one megahertz a single point ground system is preferable; above 10 MHz, a multipoint ground system is best. Between 1 and 10 MHz a single point ground can usually be used provided the length of the longest ground conductor is less than one-twentieth of a wavelength. If it is greater than one-twentieth of a wavelength a multipoint ground system should be used.

PRACTICAL LOW FREQUENCY GROUNDING

Most practical grounding systems at low frequencies are a combination of the series and parallel single point ground. Such a combination is a compromise between the need to meet the electrical noise criteria and the goal of avoiding more wiring complexity than necessary. The key to balancing these factors successfully is to group ground leads selectively, so that circuits of widely varying power and noise levels do not share the same ground return wire. Thus, several low-level circuits may share a common ground return, while other high-level circuits share a different ground return conductor.

Most systems require a minimum of three separate ground returns, as shown in Fig. 3-9. The signal ground used for low-level electronic circuits should be separated from the "noisy" ground used for circuits such as relays and motors. A third "hardware" ground should be used for

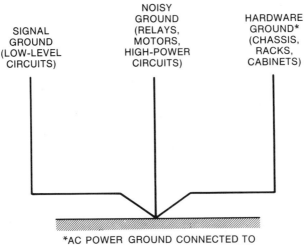

Figure 3-9. *These three classes of grounding connections should be kept separate to avoid noise coupling.*

mechanical enclosures, chassis, racks, and so on. If ac power is distributed throughout the system, the power ground (green wire) should be connected to the hardware ground. The three separate ground return circuits should be connected together at only one point. Use of this basic grounding configuration in all equipment would greatly minimize grounding problems.

An illustration of how these grounding principles might be applied to a nine-track digital tape recorder is shown in Fig. 3-10. There are three signal grounds, one noisy ground, and one hardware ground. The most sensitive circuits, the nine read amplifiers, are grounded by using two separate ground returns. Five amplifiers are connected to one, and four are connected to the other. The nine write amplifiers, which operate at a much higher level than the read amplifiers, and the interface and control logic are connected to a third ground return. The three dc motors and their control circuits, the relays, and the solenoids are connected to the noisy ground. Of these elements, the capstan motor control circuit is the most sensitive; it is properly connected closest to the primary ground point. The hardward ground provides the ground for the enclosure and housing. The signal grounds, noisy ground, and hardware ground should be connected together only at the source of primary power, that is, the power supply.

When designing the grounding system for a piece of equipment, a block diagram similar to Fig. 3-10 can be very useful in determining the proper interconnection of the various circuit grounds.

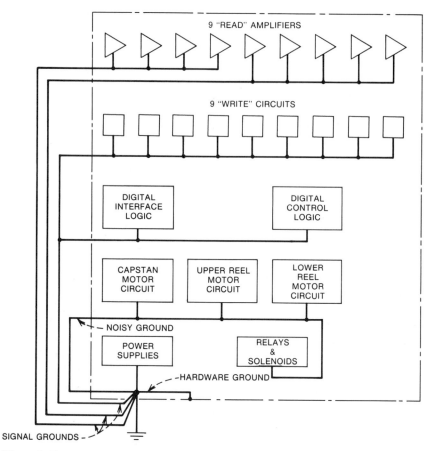

Figure 3-10. *Typical grounding system for 9-track digital tape recorder.*

HARDWARE GROUNDS

Electronic circuits for any large system are usually mounted in relay racks or cabinets. These racks and cabinets must be grounded for safety. In some systems such as electromechanical telephone offices, the racks serve as the return conductor for relay switching circuits. The rack ground is often very noisy, and it may have fairly high resistance due to joints and seams in the rack or in pull-out drawers.

Figure 3-11 shows a typical system consisting of sets of electronics mounted on panels which are then mounted to two relay racks. Rack number 1, on the left, shows correct grounding. The panel is strapped to the rack to provide a good ground and the racks are strapped together and

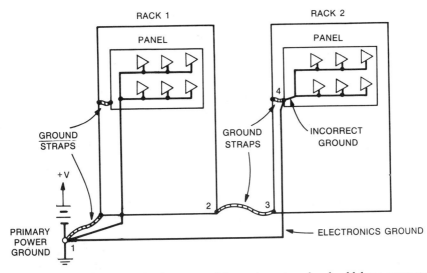

Figure 3-11. *Electronic circuits mounted in equipment racks should have separate ground connections. Rack 1 shows correct grounding, rack 2 shows incorrect grounding.*

tied to ground at the primary power source. The electronics circuit ground does not make contact with the panel or rack. In this way noise currents on the rack cannot return to ground through the electronics ground. At high frequencies, some of the rack noise current can return on the electronics ground due to capacitive coupling between the rack and electronics. This capacitance should, therefore, be kept as small as possible. Rack number 2, on the right, shows an incorrect installation in which the circuit ground is connected to the rack ground. Noise currents on the rack can now return on the electronics ground, and there is a ground loop between points 1, 2, 3, 4, 1.

If the installation does not provide a good ground connection to the rack or panel, it is best to eliminate the questionable ground, and then to provide a definite ground by some other means, or be sure that there is no ground at all. Do not depend on sliding drawers, hinges, and so on, to provide a reliable ground connection. When the ground is of a questionable nature, performance may vary from system to system or time to time, depending on whether or not the ground is made.

Hardware grounds produced by intimate contact, such as welding, brazing or soldering, are better than those made by screws and bolts. When joining dissimilar metals for grounding, care must be taken to prevent galvanic corrosion and to assure that galvanic voltages are not troublesome. Improperly made ground connections may perform perfectly well on new equipment but may be the source of mysterious trouble later.

When electrical connections are to be made to a metallic surface, such as a chassis, the metal should be protected from corrosion with a conductive coating. For example, finish aluminum with a conductive alodine or chromate finish instead of the nonconductive anodized finish. If chassis are to be used as ground planes, careful attention must also be paid to the electrical properties of seams, joints, and openings.

SINGLE GROUND REFERENCE FOR A CIRCUIT

Since two ground points are seldom at the same potential, the difference in ground potential will couple into a circuit if it is grounded at more than one point. This condition is illustrated in Fig. 3-12; a signal source is grounded at point A and an amplifier is grounded at point B. Note that in this discussion, an amplifier is generally mentioned as the load. The amplifier is simply a convenient example, however, and the grounding methods apply to any type of load. Voltage V_G represents the diffference in ground potential between points A and B. In Fig. 3-12 and subsequent illustrations, two different ground symbols are used to emphasize that two physically separated grounds are not usually at the same potential. Resistors R_{C1} and R_{C2} represent the resistance of the conductors connecting the source to the amplifier.

In Fig. 3-12 the input voltage to the amplifier is equal to $V_s + V_G$. To eliminate the noise, one of the ground connections must be removed. Elimination of the ground connection at B means the amplifier must

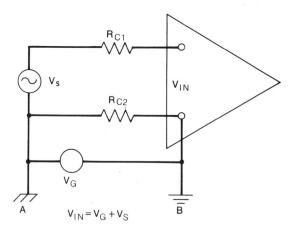

Figure 3-12. *Noise voltage V_G will couple into the amplifier if the circuit is grounded at more than one point.*

operate from an ungrounded power supply. A differential amplifier could also be used as discussed later in this chapter. It is usually easier, however, to eliminate ground connection A at the source.

The effect of isolating the source from ground can be determined by considering a low-level transducer connected to an amplifier, as shown in Fig. 3-13. Both the source and one side of the amplifier input are grounded.

For the case where $R_{C2} \ll R_s + R_{C1} + R_L$, the noise voltage V_N at the amplifier terminals is equal to

$$V_N = \left[\frac{R_L}{R_L + R_{C1} + R_s} \right] \left[\frac{R_{C2}}{R_{C2} + R_G} \right] V_G. \qquad (3\text{-}6)$$

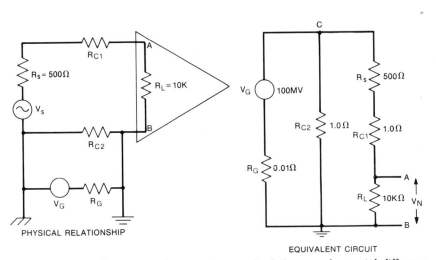

Figure 3-13. *With two ground connections, much of the ground-potential difference appears across the load as noise.*

Example 3-1. Consider the case where the ground potential in Fig. 3-13 is equal to 100 mV, a value equivalent to 10 A of ground current flowing through a ground resistance of 0.01 Ω. If $R_s = 500$ Ω, $R_{C1} = R_{C2} = 1$ Ω, and $R_L = 10$ kΩ, then from Eq. 3-6 the noise voltage at the amplifier terminals is 95 mV. Thus, almost all of the 100-mV ground differential voltage is coupled into the amplifier.

The source can be isolated from ground by adding the impedance Z_{SG}, as shown in Fig. 3-14. Ideally, the impedance Z_{SG} would be infinite, but due to leakage resistance and capacitance it has some large finite value.

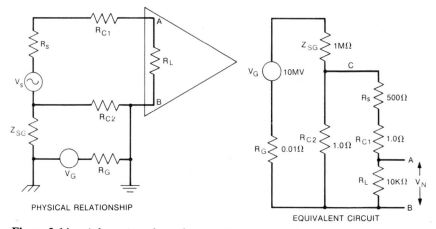

PHYSICAL RELATIONSHIP

EQUIVALENT CIRCUIT

Figure 3-14. *A large impedance between the source and ground keeps most of the ground-potential difference away from the load and reduces noise.*

For the case where $R_{C2} \ll R_s + R_{C1} + R_L$, and $Z_{SG} \gg R_{C2} + R_G$, the noise voltage V_N at the amplifier terminals is

$$V_N = \left[\frac{R_L}{R_L + R_{C1} + R_s} \right] \left[\frac{R_{C2}}{Z_{SG}} \right] V_G. \qquad (3-7)$$

Most of the noise reduction obtained by isolating the source is due to the second term of Eq. 3-7. If Z_{SG} is infinite, there is no noise voltage coupled into the amplifier. If the impedance Z_{SG} from source to ground is 1 MΩ and all other values are the same as in the previous example, the noise voltage at the amplifier terminals is, from Eq. 3-7, now only 0.095 μV. This is a reduction of 120 dB from the previous case where the source was grounded.

AMPLIFIER SHIELDS

High gain amplifiers are often enclosed in a metallic shield to provide protection from electric fields. The question then arises as to where the shield should be grounded. Figure 3-15 shows the parasitic capacitance that exists between the amplifier and the shield. From the equivalent circuit, it can be seen that the stray capacitances C_{3S} and C_{1S} provide a feedback path from output to input. If this feedback is not eliminated, the amplifier may oscillate. *The only shield connection that will eliminate the unwanted feedback path is the one shown at the bottom of Fig. 3-15 where the shield is connected to the amplifier common terminal.* By connecting the

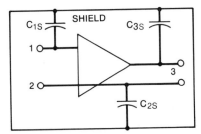

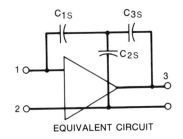

PHYSICAL RELATIONSHIP

EQUIVALENT CIRCUIT

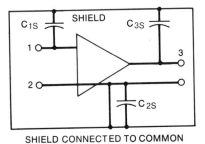

SHIELD CONNECTED TO COMMON

Figure 3-15. *Amplifier shield should be connected to the amplifier common.*

shield to the amplifier common, capacitance C_{2S} is short circuited and the feedback is eliminated. This shield connection should be made even if the common is not at earth ground.

GROUNDING OF CABLE SHIELDS

Shields on cables used for low frequency signals should be grounded at only one point when the signal circuit has a single point ground. If the shield is grounded at more than one point, noise current will flow. In the case of a shielded twisted pair, the shield currents may inductively couple unequal voltages into the signal cable and be a source of noise. In the case of coaxial cable, the shield current generates a noise voltage by causing an IR drop in the shield resistance, as was shown in Fig. 2-20. But if the shield is to be grounded at only one point, where should that point be? The top drawing in Fig. 3-16 shows an amplifier and the input signal leads with an ungrounded source. Generator V_{G1} represents the potential of the amplifier common terminal above earth ground, and generator V_{G2} represents the difference in ground potential between the two ground points.

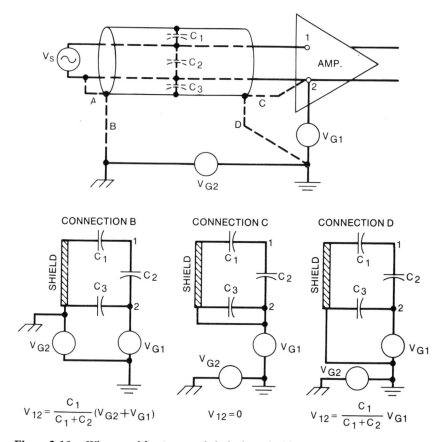

Figure 3-16. *When amplifier is grounded, the best shield connection is (C), with shield connected to amplifier common.*

Since the shield has only one ground, it is the capacitance between the input leads and the shield that provides the noise coupling. The input shield may be grounded at any one of four possible points through the dotted connections labeled *A*, *B*, *C*, and *D*. Connection *A* is obviously not desirable, since it allows shield noise current to flow in one of the signal leads. This noise current flowing through the impedance of the signal lead produces a noise voltage in series with the signal.

The three lower drawings in Fig. 3-16 are equivalent circuits for grounding connections *B*, *C*, and *D*. Any extraneous voltage generated between the amplifier input terminals (points 1 and 2) is a noise voltage. With grounding arrangement *B*, a voltage is generated across the amplifier input terminals due to the generators V_{G2} and V_{G1} and the capacitive voltage divider formed by C_1 and C_2. This connection, too, is unsatisfactory. For ground connection C, there is no voltage V_{12} regardless of the value of generators V_{G1} or V_{G2}. With ground connection D, a voltage is generated across the amplifier input terminals due to generator V_{G1} and the capacitive voltage divider C_1 and C_2. The only connection which precludes a noise voltage V_{12} is connection C. Thus, *for a circuit with an ungrounded source and a grounded amplifier, the input shield should always be connected to the amplifier common terminal, even if this point is not at earth ground.*

The case of an ungrounded amplifier connected to a grounded source is shown in Fig. 3-17. Generator V_{G1} represents the potential of the source common terminal above the actual ground at its location. The four

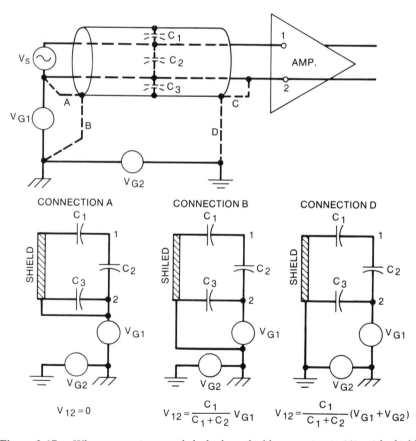

Figure 3-17. *When source is grounded, the best shield connection is (A), with shield connected to the source common. The configuration can also be used with a differential amplifier.*

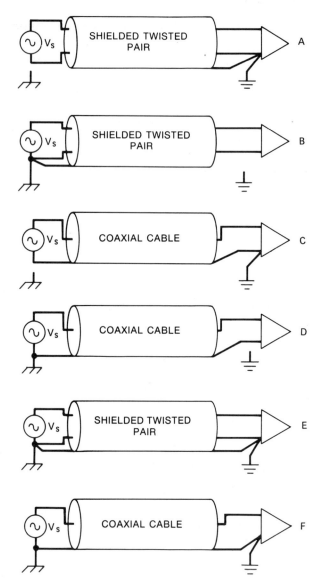

Figure 3-18. *Preferred grounded schemes for shielded, twisted pairs and coaxial cable at low frequency.*

possible connections for the input cable shield are again shown as the dashed lines labeled *A*, *B*, *C*, and *D*. Connection *C* is obviously not desirable since it allows shield noise currents to flow in one of the signal conductors in order to reach ground. Equivalent circuits are shown at the bottom of Fig. 3-17 for shield connections *A*, *B*, and *D*. As can be seen, only connection *A* produces no noise voltage between the amplifier input terminals. *Therefore, for the case of a grounded source and ungrounded amplifier, the input should be connected to the source common terminal, even if this point is not at earth ground.*

Preferred low-frequency shield grounding schemes for both shielded twisted pair and coaxial cable are shown in Fig. 3-18. Circuits *A* through *D* are grounded at the amplifier or the source, but not at both ends.

When the signal circuit is grounded at both ends, the amount of noise reduction possible is limited by the difference in ground potential and the susceptibility of the ground loop to magnetic fields. The preferred shield ground configurations for cases where the signal circuit is grounded at both ends are shown in circuits *E* and *F* of Fig. 3-18. In circuit *F*, the shield of the coaxial cable is grounded at both ends to force some ground-loop current to flow through the lower-impedance shield, rather than the center conductor. In the case of circuit *E* the shielded twisted pair is also grounded at both ends to shunt some of the ground-loop current from the signal conductors. If additional noise immunity is required, the ground loop must be broken. This can be done by using transformers, optical couplers, or a differential amplifier.

An indication of the type of performance to be expected from the configurations shown in Fig. 3-18 can be obtained by referring to the results of the magnetic coupling experiment presented in Figs. 2-22 and 2-23.

ISOLATION AND NEUTRALIZING TRANSFORMERS

A ground loop is formed when both ends of a circuit are grounded, as shown in Fig. 3-19. The loop can be broken by using an isolation transformer as shown in the lower drawing. In some circuits, however, direct current or very low frequency continuity is required between the two circuits and an isolation transformer is not practical. Under these conditions, a transformer can be used as a longitudinal choke (also called a neutralizing transformer or balun), as shown in Fig. 3-20. A transformer connected in this manner presents a low impedance to the signal current and allows dc coupling. To any longitudinal (common mode) noise current, however, the transformer is a high impedance.

The signal current shown in Fig. 3-20 flows equally in the two conductors but in opposite directions. This is the desired signal, and it is also

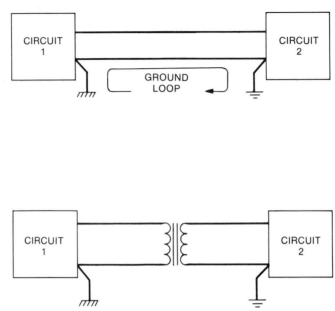

Figure 3-19. *A ground loop between two circuits can be broken by inserting an isolation transformer.*

known as the differential circuit current or metallic circuit current. The noise currents flow in the same direction along both conductors and are called longitudinal currents or common mode currents.

Circuit performance for Fig. 3-20 may be analyzed by referring to the equivalent circuit. Voltage generator V_s represents a signal voltage that is connected to the load R_L by conductors with resistance R_{C1} and R_{C2}. The neutralizing transformer is represented by the two inductors L_1 and L_2 and the mutual inductance M. If both windings of the transformer are identical and closely coupled on the same core, then L_1, L_2, and M are equal. Voltage generator V_G represents a longitudinal voltage due either to magnetic coupling in the ground loop or to a ground differential voltage. Since the conductor resistance R_{C1} is in series with R_L and of much smaller magnitude, it can be neglected.

The first step is to determine the response of the circuit to the signal voltage V_s, neglecting the effect of V_G. The circuit of Fig. 3-20 can be redrawn, as shown in Fig. 3-21. This figure is identical to the circuit of Fig. 2-17. There it was shown that at frequencies greater than $\omega = 5R_{C2}/L_2$, virtually all the current I_s returned to the source through the second conductor and not through the ground plane. If L_2 is chosen such that the lowest signal frequency is greater than $5R_{C2}/L_2$ radians per second, then

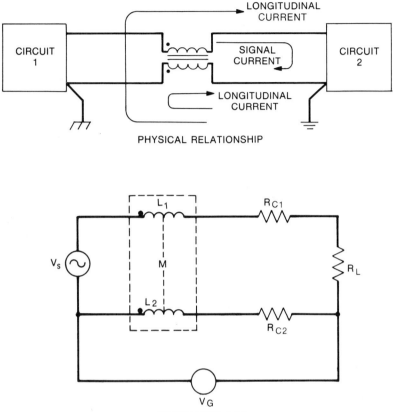

Figure 3-20. *When dc or low frequency continuity is required, a longitudinal choke can be used to break a ground loop.*

$I_G = 0$. Under these conditions, the voltages around the top loop of Fig. 3-21 can be summed as follows:

$$V_s = j\omega(L_1 + L_2)I_s - 2j\omega MI_s + (R_L + R_{C2})I_s. \qquad (3\text{-}8)$$

Remembering that $L_1 = L_2 = M$ and solving for I_s gives

$$I_s = \frac{V_s}{R_L + R_{C2}} = \frac{V_s}{R_L}, \qquad (3\text{-}9)$$

provided R_L is much greater than R_{C2}. Equation 3-9 is the same that would have been obtained if the transformer had not been present. It therefore,

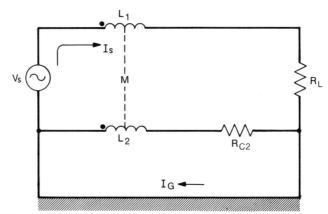

Figure 3-21. *Equivalent circuit for Fig. 3-20 for analysis of response to signal voltage* V_s.

has no effect on the signal transmission so long as the transformer inductance is large enough that the signal frequency ω is greater than $5R_{C2}/L_2$.

The response of the circuit of Fig. 3-20 to the longitudinal voltage V_G can be determined by considering the equivalent circuit shown in Fig. 3-22. If the transformer were not present, the complete noise voltage V_G would appear across R_L.

When the transformer is present, the noise voltage developed across R_L can be determined by writing equations around the two loops shown in the

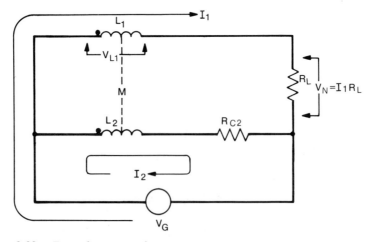

Figure 3-22. *Equivalent circuit for Fig. 3-20 for analysis of response to longitudinal voltage* V_G.

illustration. Summing voltages around the ouside loop gives

$$V_G = j\omega L_1 I_1 + j\omega M I_2 + I_1 R_L. \qquad (3\text{-}10)$$

The sum of the voltages around the lower loop is

$$V_G = j\omega L_2 I_2 + j\omega M I_1 + R_{C2} I_2. \qquad (3\text{-}11)$$

Equation 3-11 can be solved for I_2, giving the following result:

$$I_2 = \frac{V_G - j\omega M I_1}{j\omega L_2 + R_{C2}}. \qquad (3\text{-}12)$$

Remembering that $L_1 = L_2 = M = L$, and substituting Eq. 3-12 into Eq. 3-10, and solving for I_1, gives

$$I_1 = \frac{V_G R_{C2}}{j\omega L (R_{C2} + R_L) + R_{C2} R_L}. \qquad (3\text{-}13)$$

The noise voltage V_N is equal to $I_1 R_L$, and since R_{C2} is normally much less than R_L we can write

$$V_N = \frac{V_G R_{C2}/L}{j\omega + R_{C2}/L}. \qquad (3\text{-}14)$$

A plot of V_N/V_G is shown in Fig. 3-23. To minimize this noise voltage, R_{C2} should be kept as small as possible and the transformer inductance L should be such that

$$L \gg \frac{R_{C2}}{\omega} \qquad (3\text{-}15)$$

where ω is the frequency of the noise. The transformer also must be large

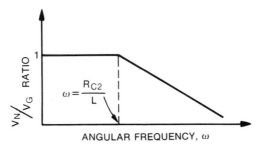

Figure 3-23. *Noise voltage may be significant if R_{C2} is large.*

enough that any unbalanced dc currents flowing in the circuit does not cause saturation.

The longitudinal choke or neutralizing transformer shown in Fig. 3-20 can be easily made; simply wind the conductors connecting the two circuits around a magnetic core, as shown in Fig. 3-24. The signal conductors from more than one circuit may be wound around the same core without the signal circuits interfering (crosstalking). In this way one core can be used to provide a neutralizing transformer for many circuits. A typical neutralizing transformer used in the telephone plant will contain 25–50 circuits.

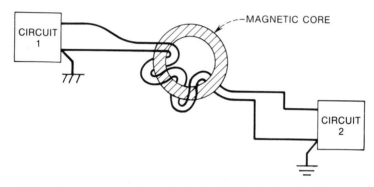

Figure 3-24. *An easy way to place a logitudinal choke in the circuit is to wind both conductors around a toroidal magnetic core. A coaxial cable may also be used in place of the conductors shown.*

OPTICAL COUPLERS

Another way to break the ground loop between two circuits is to use an optical coupler, as shown in Fig. 3-25. The basic optical coupler consists of a light emitting diode (LED) optically coupled to a transistor, diode, or thyristor. Both devices are contained in the same package. This type of

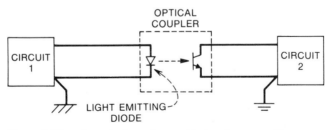

Figure 3-25. *An optical coupler used to break a ground loop.*

circuit gives almost perfect isolation against any difference in ground potential, because the only connection between circuits 1 and 2 is through the light beam of the optical coupler.

Optical couplers are especially useful in digital circuits. For analog circuits, however, they are less suitable because linearity through the coupler is not always satisfactory. Recently, however, analog circuits have been designed using optical feedback techniques to compensate for the inherent nonlinearity of the coupler (Waaben, 1975).

DIFFERENTIAL AMPLIFIERS

A differential (or balanced-input) amplifier may be used to decrease the effect of a longitudinal (common-mode) noise voltage. This is shown in the upper drawing of Fig. 3-26, where V_G is the longitudinal voltage. The differential amplifier has two input voltages V_1 and V_2, and the output

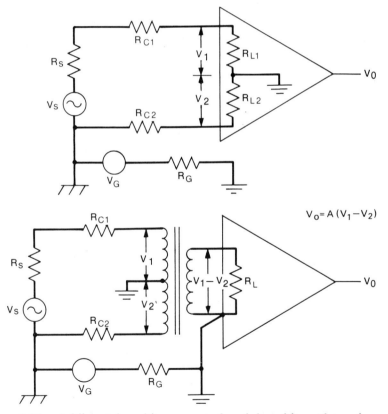

Figure 3-26. *A differential amplifier—or a single-ended amplifier with transformer— can be used to reduce the effects of a common-mode noise voltage.*

voltage is equal to the amplifier gain (A) times the difference in the two input voltages, $V_0 = A(V_1 - V_2)$.

The lower drawing of Fig. 3-26 shows how a single-ended (or unbalanced) amplifier can be used to simulate the performance of a true balanced amplifier. The transformer primary has a grounded center tap, and the voltages across the two halves are V_1 and V_2. The secondary voltage (assuming a $1:1$ turns ratio) is equal to $V_1 - V_2$. Amplifier output again is equal to the gain times this voltage difference, duplicating the balanced amplifier output.

The response of either circuit in Fig. 3-26 to the noise voltage can be determined from the equivalent circuit shown in Fig. 3-27. For resistance R_{L2} much larger than R_G, the input voltage to the amplifier due to common-mode noise voltage V_G is as follows:

$$V_N = V_1 - V_2 = \left(\frac{R_{L1}}{R_{L1} + R_{C1} + R_s} - \frac{R_{L2}}{R_{L2} + R_{C2}} \right) V_G. \qquad (3\text{-}16)$$

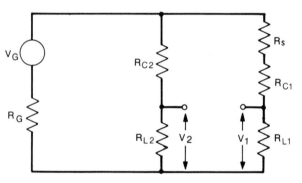

EQUIVALENT CIRCUIT

Figure 3-27. *Equivalent circuit for analysis of differential-amplifier circuit.*

Example 3-2. If in Fig. 3-27, $V_G = 100$ mV, $R_G = .01$ Ω, $R_s = 500$ Ω, $R_{C1} = R_{C2} = 1$ Ω, and $R_{L1} = R_{L2} = 10$ kΩ, then from Eq. 3-16, $V_N = 4.6$ mV. If, however, R_{L1} and R_{L2} were 100 kΩ instead of 10 kΩ, then $V_N = 0.5$ mV. This represents an almost 20 dB decrease in the input noise voltage.

From the above example, it is obvious that increasing the input impedance (R_{L1} and R_{L2}) of the differential amplifier decreases the noise voltage coupled into the amplifier due to V_G. From Eq. 3-16, it can be seen that decreasing the source resistance R_s also decreases the noise voltage coupled into the amplifier. Figure 3-28 shows a way to modify the circuits of Fig. 3-26 to increase the input impedance of the amplifier to the

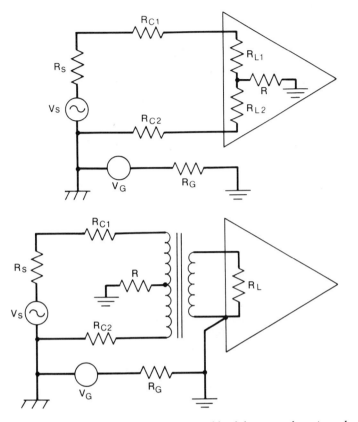

Figure 3-28. *Insertion of resistance R into ground lead decreases the noise voltage.*

longitudinal voltage V_G without increasing the input impedance to the signal voltage V_s. This is done by adding resistor R into the ground lead as shown. When using a high-impedance differential amplifier, both the input cable shield and the source common should be grounded at the source as was shown in Fig. 3.18 *B*.

SHIELD GROUNDING AT HIGH FREQUENCIES

At frequencies less than 1 MHz, shields should normally be grounded at one end only. Otherwise, as previously explained, large power-frequency currents can flow in the shield and introduce noise into the signal circuit. The single point ground also eliminates the shield ground loop and its associated magnetic pickup.

At frequencies above 1 MHz or where cable length exceeds one-twentieth of a wavelength it is often necessary to ground a shield at more than one point to guarantee that it remains at ground potential. Another problem develops at high frequencies; stray capacitive coupling tends to complete the ground loop, as shown in Fig. 3-29. This makes it difficult or impossible to maintain isolation at the ungrounded end of the shield.

It is therefore common practice at high frequencies to ground cable shields at both ends. For long cables, ground may be required every one-tenth of a wavelength. The noise voltage due to a difference in ground potential that couples into the circuit (primarily at power frequencies and its harmonics) can usually be filtered out, because there is a large frequency difference between the noise and the signal frequency. At frequencies above one megahertz the skin effect reduces the coupling due to signal and noise current flowing on the shield. This skin effect causes the noise current to flow on the outside surface of the shield and the signal current to flow on the inside surface of the shield. The multiple ground also provides a degree of magnetic shielding at higher frequencies when coaxial cable is used.

The characteristics of the circuit shown in Fig. 3-29 can be put to advantage by replacing the stray capacitance with a small capacitor, thus forming a combination or hybrid ground. At low frequencies a single point ground exists since the impedance of the capacitor is large. However, at high frequencies the capacitor becomes a low impedance, thus converting the circuit to one having a multiple ground. Such a ground configuration is often useful for circuits that must operate over a very wide frequency range.

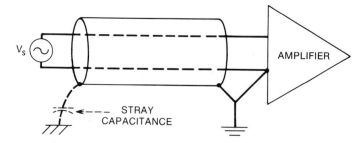

Figure 3-29. *At high frequencies, stray capacitance tends to complete the ground loop.*

GUARD SHIELDS

Noise reduction greater than that obtainable with a differential amplifier can be obtained by using an amplifier with a guard shield. A guard shield is placed around the amplifier and held at a potential which prevents current flow in the unbalanced source impedance. The effect of a guard shield can best be explained by considering an example in which a guard shield is used to cancel the effects of a difference in ground potential.

Figure 3-30 shows an amplifier connected by a shielded twisted pair to a grounded source. V_G is a common-mode (longitudinal) voltage due to a difference in ground potentials. V_s and R_s are the differential signal voltage and source resistance, respectively. R_{IN} is the input impedance to the amplifier. C_{1G} and C_{2G} are stray capacitances between the amplifier input terminals and ground, including the cable capacitance. There are two undesirable currents flowing as a result of voltage V_G. Current I_1 flows through resistors R_s and R_1, and capacitance C_{1G}. Current I_2 flows through resistor R_2 and C_{2G}. If each current does not flow through the same total impedance there will be a differential input voltage to the amplifier. If, however, a guard shield is placed around the amplifier, as shown in Fig. 3-31, and the shield is held at the same potential as point A, currents I_1 and I_2 both become zero because both ends of the path are at the same potential. Capacitances C_1 and C_2 now appear between the input terminals and the shield.

The shield accomplishes the objective of eliminating the differential input noise voltage. Unmentioned, however, has been the problem of how to hold the shield at the potential of point A. One way to do this is shown in Fig. 3-32, where the guard shield is held at the potential of point A by connecting it to the cable shield. The other end of the cable shield is then grounded at point A. This discussion assumes that the source common (lower) terminal is at the same potential as point A. That is, there is no noise voltage generated between point A and the source common. If there is any possibility of a noise voltage being generated between the common

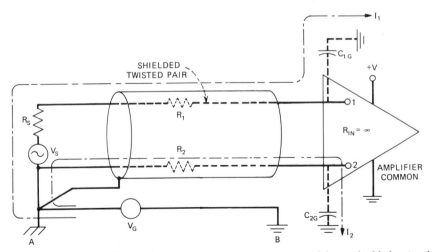

Figure 3-30. *Amplifier and a grounded source are connected by a shielded twisted pair.*

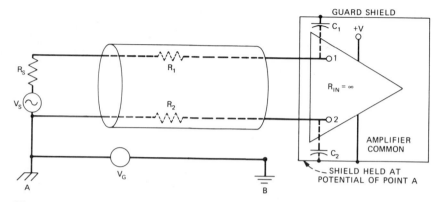

Figure 3-31. *Guard shield at potential of point A eliminates noise currents.*

terminal of V_s and ground point A, the guard shield should be connected to the source common as illustrated, instead of directly to point A.

Notice that the amplifier and shield connections of Fig. 3-32 do not violate any of the previously described rules. The cable shield is grounded at only one point (point A). The input cable shield is connected to the amplifier common. The shield around the amplifier is also connected to the amplifier common terminal.

In the guarded amplifier of Fig. 3-32, any ground point at potential B inside the amplifier guard shield increases the capacitance from the input leads to ground (unguarded capacitance). For the scheme to work, therefore, it means the amplifier must be powered by self contained batteries, or else power must be brought in through an electrostatically shielded transformer. No point of the guard shield can come in contact with ground B without nullifying its effectiveness. A practical circuit, therefore, has a

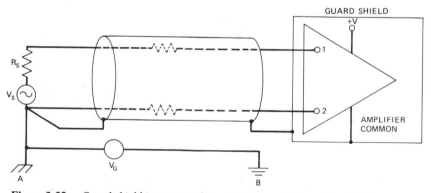

Figure 3-32. *Guard shield is connected to point A through the cable shield.*

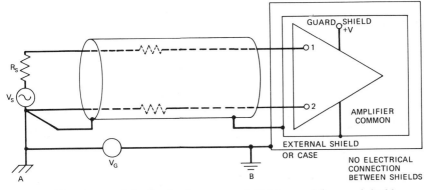

Figure 3-33. *Practical circuit often has a second shield around the guard shield.*

second shield placed around the guard shield to guarantee the guard's integrity, as shown in Fig. 3-33. This second or external shield is grounded to the local ground, point *B*, and satisfies the safety requirements.

A guard shield is usually only required when extremely-low-level signals are being measured, or when very large common-mode voltages are present and all other noise reduction techniques have also been applied to reduce the noise pick-up to an absolute minimum. A guard shield may be used around a single-ended amplifier as well as a differential amplifier.

Example 3-3. Consider a numerical example, as illustrated in Fig. 3-34, where $R_1 = R_2 = 0$, $R_s = 2.6$ kΩ, $C_{1G} = C_{2G} = 100$ pF and $V_G = 100$ mV at 60 Hz. The reactance of 100 pF is 26 MΩ at 60 Hz. The differential input

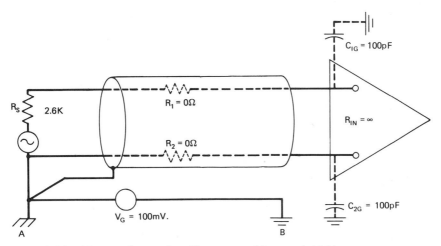

Figure 3-34. *Numerical example to illustrate need for guard shield.*

noise voltage across the amplifier input terminals without a guard shield can be written as

$$V_N = \left(\frac{R_s + R_1}{R_s + R_1 + Z_{1G}} - \frac{R_2}{R_2 + Z_{2G}} \right) V_G, \qquad (3\text{-}17)$$

where Z_{1G} and Z_{2G} are the impedance of capacitance C_{1G} and C_{2G}, respectively. Substituting numerical values into Eq. 3-17, the input noise voltage without the guard shield is 10 μV. If the use of the guard shield reduces each line's capacitance to ground to 2 pF, as shown in Fig. 3-35, the differential input noise voltage across the amplifier input terminals with the guard shield in place can still be written as shown in Eq. 3-17, but the input noise voltage is now reduced to 0.2 μV, a 34 dB improvement. The 2-pF capacitance to ground is due to the fact that the guard shield is not perfect. If it were perfect, there would be no capacitance to ground and the noise voltage would be zero. It should be noted that the noise voltage coupled into the amplifier increases as the frequency of the noise source is increased, since the impedance of C_{1G} and C_{2G} decrease as the frequency is increased.

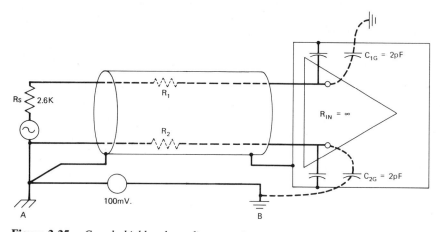

Figure 3-35. *Guard shield reduces line capacitance to ground and, therefore, noise voltage.*

GUARDED METERS

Even for those who do not intend to design equipment using a guard shield, there is still a good reason to understand the operating principles. Many new measuring instruments are being manufactured with a guard shield (see Fig. 3-36). It is up to the user to connect the guard shield to the

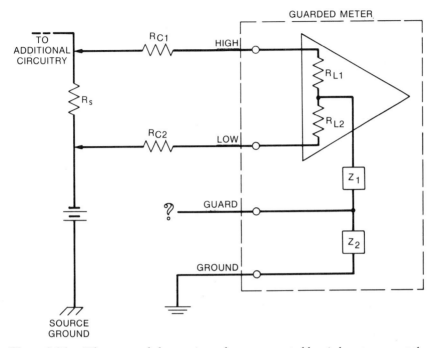

Figure 3-36. *When a guarded meter is used, a common problem is how to connect the guard terminal.*

proper place in the circuit being measured. When a user does not understand the purpose of a guard shield, he is likely to leave it open or connect it to the meter ground; neither of these connections produces optimum results. To take maximum advantage of the guard shield, the following rule should be followed: *The guard shield should always be connected such that no common-mode current can flow through any of the input resistances.* This normally means connecting the guard to the low-impedance terminal of the source.

Example 3-4. Refer to Fig. 3-36. The problem is to measure the voltage across resistor R_s, neither end of which is grounded, with a guarded digital voltmeter. What is the best connection for the guard shield? Five possible ways to connect the guard shield are shown in Figs. 3-37 through 3-41. Voltage V_G is the ground differential voltage and V_N is the battery noise voltage. Figure 3-37 shows the best connection, with the guard connected to the low-impedance terminal of the source. Under this condition no noise current flows through the input circuit of the meter.

The connection shown in Fig. 3-38, where the guard is connected to ground at the source is not as good as the previous connection. Here, the noise current from the generator V_G is no problem, but noise current from V_N flows through impedances R_{C2}, R_{L2}, and Z_1, and causes a noise voltage to be coupled into the amplifier. The connections of Figs. 3-39, 3-40, and 3-41 all allow noise current to flow through the meter input circuit and are, therefore, undesirable.

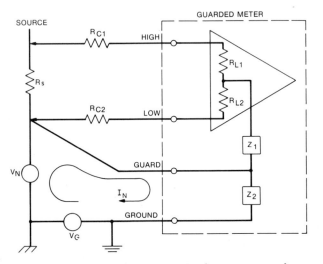

Figure 3-37. *When measuring voltage across R_s, best connection for guard is to the low-impedance side of R_s; noise current does not affect amplifier.*

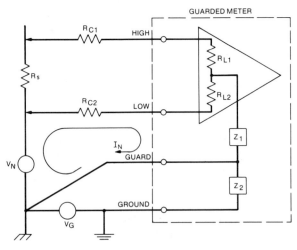

PROTECTION AGAINST V_G ONLY.

Figure 3-38. *Guard connected to source ground gives no protection against V_N.*

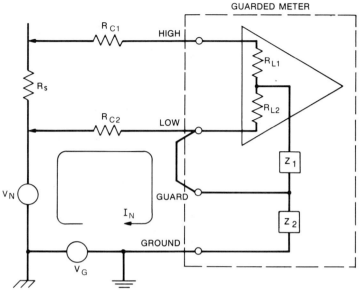

NOISE VOLTAGE APPEARS ACROSS R $_{C2}$

Figure 3-39. *Guard connected to low side of meter allows noise current to flow in line resistance R_{C2}.*

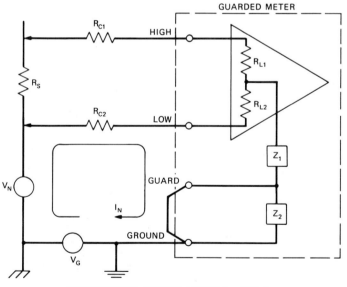

NOISE VOLTAGE APPEARS ACROSS R$_{C2}$ AND R$_{L2}$

Figure 3-40. *Guard connected to local ground is ineffective; noise current flows through R_{C2}, R_{L2}, and Z_1.*

87

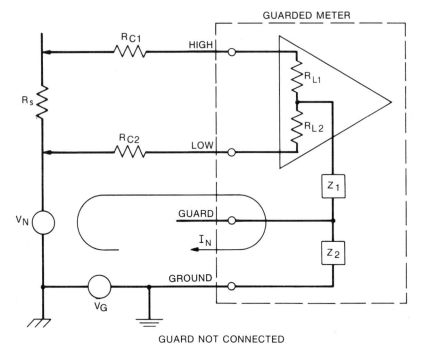

Figure 3-41. *Guard not connected; noise currents due to V_N and V_G flow through R_{C2}, R_{L2}, Z_1, and Z_2.*

CABLES AND CONNECTORS

Inadvertent generation of ground loops and poor shielding practices are likely to occur in system cabling, especially when different design groups are responsible for different sides of the interface. Good cabling requires design; it is not something that just happens.

When possible, high-level and low-level leads should not be put in the same cable. If they must be in one cable, the high-level leads should be grouped and put in a shield. Normal precautions should be taken with the low-level leads.

Low- and high-level leads should be run through separate connectors where possible. If high- and low-level leads must be in one connector, they should be placed on pins which are physically separated. Ground leads should be placed on the intervening pins, as shown in Fig. 3-42. If all pins in the connector are not used, the spare one(s) should be in the middle, separating the high- and low-level leads.

Shielding integrity should be maintained when cables are run between systems. Cable shields should be carried through connectors. When more

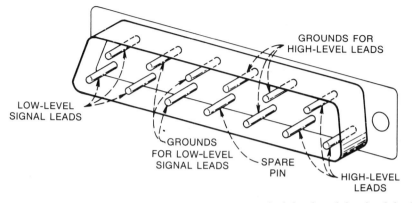

Figure 3-42. *When assigning pins in a connector, high-level and low-level leads should be physically separated with ground leads in between.*

than one shielded cable goes through a connector, each shield should be carried through on a separate pin. Connecting all the shields to a single pin produces ground loops and allows shield currents to flow between individual shields.

Where low-level signal cables require shielding and are grounded at one point only, insulation is necessary over the shield. This prevents the shield from inadvertently touching ground at some other point.

SUMMARY

- At low frequencies, a single point ground system should be used.
- At high frequencies, a multipoint ground system should be used.
- A low frequency system should have a minimum of three separate ground returns. These should be
     signal ground,
     noisy ground,
     hardward ground.
- The basic objectives of a good ground system are
     minimize noise voltage from two ground currents flowing through a common impedance,
     avoid generating ground loops.
- For the case of a grounded amplifier with an ungrounded source, the input cable shield should be connected to the amplifier common terminal.
- For the case of a grounded source with an ungrounded amplifier, the input cable shield should be connected to the source common terminal.
- A shield around a high-gain amplifier should be connected to the amplifier common.

- When a signal circuit is grounded at both ends, the ground loop formed is susceptible to noise from
    magnetic fields,
    differential ground voltages.
- Methods of breaking ground loops are
    isolation transformer,
    neutralizing transformer,
    optical couplers,
    differential amplifiers,
    guarded amplifiers.
- At high frequencies, shields around signal cables are usually grounded at more than one point.

BIBLIOGRAPHY

Ady, R., "Applying Opto-Isolators," *Electronic Products*, June 17, 1974.

Bell Laboratories, *Physical Design of Electronic Systems*, Vol. 1, Chapter 10 (Electrical Interference); Prentice-Hall, Englewood Cliffs, N.J., 1970.

Brown, H., "Don't Leave System Grounding to Chance," *EDN/EEE*, January 15, 1972.

Buchman, A. S., "Noise Control in Low Level Data Systems," *Electromechanical Design*, September, 1962.

Cushman, R. H., "Designer's Guide to Optical Couplers," *EDN*, July 20, 1973.

Ficchi, R. O., *Electrical Interference*, Hayden Book Co., New York 1964.

Ficchi, R. O., *Practical Design For Electromagnetic Compatability*, Hayden Book Co., New York 1971.

Frederick Research Corp., *Handbook on Radio Frequency Interference*, Vol. 3 (Methods of Electromagnetic Interference Suppression) Frederick Research Corp., Wheaton, Maryland 1962.

Hewlett-Packard, *Floating Measurements and Grounding,* application note 123, 1970.

Morrison, R., *Grounding and Shielding Techniques in Instrumentation*, Wiley, New York, 1967.

Nalle, D., "Elimination of Noise in Low Level Circuits," *ISA Journal*, Vol. 12, August, 1965.

National Electrical Code, National Fire Protection Association, Boston, Mass., 1975. (This code is normally reissued every three years.)

Waaben, S., "High Performance Optocoupler Circuits," International Solid-State Circuits Conference, Philadelphia, Pa., February, 1975.

White, D. R. J., *Electromagnetic Interference and Compatibility*, Vol. 3 (EMI Control Methods and Techniques), Don White Consultants, Germantown, Maryland 1973.

4 OTHER NOISE REDUCTION TECHNIQUES

BALANCING

A balanced circuit is a two-conductor circuit in which both conductors and all circuits connected to them have the same impedance with respect to ground and to all other conductors. The purpose of balancing is to make the noise pickup equal in both conductors, in which case it will be a longitudinal or common mode signal which can be made to cancel out in the load. Balancing is a noise reduction technique which may be used in conjunction with shielding when noise must be reduced below the level obtainable with shielding alone. In addition it is used, in some applications, in place of shielding as the primary noise reduction technique.

The use of a differential amplifier, as previously shown in Fig. 3-26, was the first step toward a balanced system. The amplifier provided a balanced load, but the source was still unbalanced due to the resistance R_s. Balancing the source with respect to ground completely balances the system, as shown in Fig. 4-1. In the general case, two common-mode noise voltages

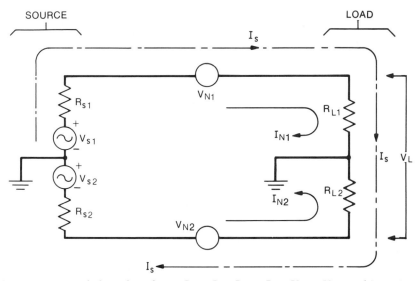

Figure 4-1. *For balanced condition: $R_{s1} = R_{s2}$, $R_{L1} = R_{L2}$, $V_{N1} = V_{N2}$, and $I_{N1} = I_{N2}$.*

V_{N1} and V_{N2} are shown in series with the conductors. These noise voltages produce noise currents I_{N1} and I_{N2}. The sources V_{s1} and V_{s2} together produce the signal current I_s. The total voltage V_L developed across the load is equal to the following:

$$V_L = I_{N1}R_{L1} - I_{N2}R_{L2} + I_s(R_{L1} + R_{L2}). \tag{4-1}$$

The first two terms represent noise voltages and the third term represents the desired signal voltage. If I_{N1} is equal to I_{N2} and R_{L1} is equal to R_{L2}, then the noise voltage across the load is equal to zero. Equation 4-1 then reduces to

$$V_L = I_s(R_{L1} + R_{L2}), \tag{4-2}$$

which represents a voltage due only to the signal current I_s.

In the balanced circuit shown in Fig. 4-2, V_1 and V_2 represent inductive pickup voltages. Generator V_3 is a noise voltage that is capacitively coupled into the circuit through C_{31} and C_{32}. The difference in ground potential betwen source and load is represented by V_G. The noise voltage produced between load terminals 1 and 2, due to voltage V_3, can be determined by referring to Fig. 4-3. Impedances Z_1 and Z_2 represent the total impedance to ground from conductors 1 and 2, respectively.

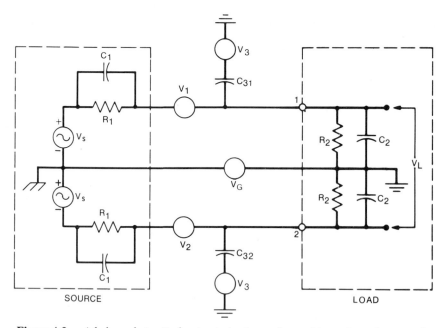

Figure 4-2. *A balanced circuit showing inductive and capacitive noise voltages and a difference in ground potential between source and load.*

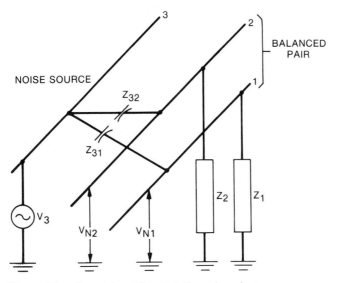

Figure 4-3. *Capacitive pickup in balanced conductors.*

For capacitive coupling the noise voltage induced into conductor 1 due to the voltage V_3 is

$$V_{N1} = \left(\frac{Z_1}{Z_1 + Z_{31}} \right) V_3. \qquad (4\text{-}3)$$

The noise voltage induced into conductor 2 due to voltage V_3 is

$$V_{N2} = \left(\frac{Z_2}{Z_2 + Z_{32}} \right) V_3. \qquad (4\text{-}4)$$

If the circuit is balanced, impedances Z_1 and Z_2 are equal. If conductors 1 and 2 are a twisted pair, impedance Z_{31} should nearly equal Z_{32}. Under these conditions V_{N1} approximately equals V_{N2} and the capacitively coupled noise voltages cancel at the load. A twisted pair can, therefore, provide protection against capacitive coupling if the circuit is balanced. Since a twisted pair can also protect against magnetic fields, a balanced circuit using a twisted pair can protect against both magnetic and electric fields without a shield over the conductors. Shields are still desirable, however, since it is difficult to obtain perfect balance and, hence, additional protection may be required.

Twisted pairs or shielded twisted pairs are usually used as the conductors in a balanced circuit, since a twisted pair is inherently a balanced configuration. A coaxial cable, on the other hand, is inherently an unbalanced configuration. If coaxial cable is to be used in a balanced system, two cables should be used, as shown in Fig. 4-4.

It should be noticed in Fig. 4-2 that the difference in ground potential (V_G) between source and load produces equal voltages at terminals 1 and 2 of the load. These voltages cancel, producing no new noise voltage across the load.

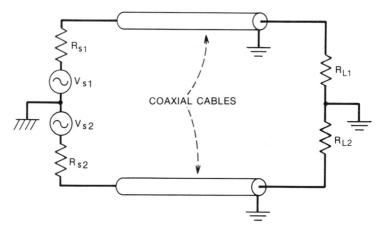

Figure 4-4.   *Use of coaxial cables in a balanced circuit.*

The degree of balance, or common mode rejection ratio (CMRR), is defined as the ratio of the common mode (or longitudinal) noise voltage to the differential (or metallic) noise voltage produced by it. It is usually expressed in decibels (dB).* This conversion from common-mode to differential voltage results from the unbalances present in the system. Referring to Fig. 4-5, the balance (or CMRR) of the circuit (in dB) is

$$\text{Balance} = 20 \log \left(\frac{V_N}{V_M} \right) \quad \text{dB.} \qquad (4\text{-}5)$$

If the source resistances R_s are small compared to the load R_L, then the common mode voltage V_C equals V_N, and V_C can be used in Eq. 4-6 in place of V_N. Then,

$$\text{Balance} = 20 \log \left(\frac{V_C}{V_M} \right) \quad \text{dB.} \qquad (4\text{-}6)$$

*See Appendix A for discussion of the decibel.

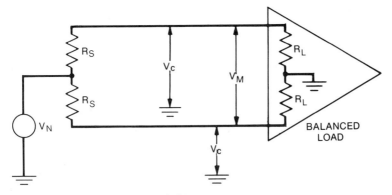

Figure 4-5. *Measuring circuit balance.*

If the source and load in Fig. 4-5 are physically separated by an appreciable distance, the balance defined in Eq. 4-6 is normally used because both measurements can be made at the same end of the circuit.

The better the balance, the greater the noise reduction obtainable. If the balance could be made perfect, no noise could enter the system. Typically, 60–80 dB of balance is reasonable to expect from a well designed circuit. Balance better than this range is possible, but special cables are usually required and individual circuit trimming may be necessary.

System balance is dependent on source balance, signal lead balance, and load balance, as well as the balance of any stray or parasitic impedances. Both resistive and reactive balances must be obtained between the two input conductors. Therefore, the resistances and reactances of each conductor to ground must be equal. The magnitude of any noise coupled into a balanced circuit is a function of the degree of unbalance and is directly proportional to the common mode noise voltage. Balance is never perfect, and some noise voltage couples into the circuit whenever common mode noise voltages are present. The common mode noise voltage can be decreased by proper shielding and grounding, as discussed in the previous chapters, and by eliminating the ground at one end of the circuit.

Example 4-1. A circuit is built with 60 dB of balance. The cables are not shielded, and each cable picks up a common mode voltage of approximately 300 mV due to electric field coupling. The noise coupled into the load is 60 dB below this or 300 μV. If a grounded shield is now placed around the conductors, the common mode pick up voltage is reduced to 13 mV. The noise coupled into the amplifier is 60 dB below that, or 13 μV. This example shows that the effects of shielding and balancing are additive. The shielding can be used to reduce the amount of common mode voltage coupled into the conductors and balancing reduces that portion of the common mode voltage which is coupled into the load.

Circuit balance also depends on the operating frequency. Normally, the higher the frequency, the harder it is to obtain good balance, because stray capacitance has more effect on circuit balance at high frequency.

Knowing the balance provided by the individual components* that make up a system does not allow prediction of overall system balance when the components are combined. For example, the unbalances in two of the components may complement each other such that the combined balance is greater than that of either of the individual components. On the other hand, the components may be such that the combined balance is less than that of either of the individual components.

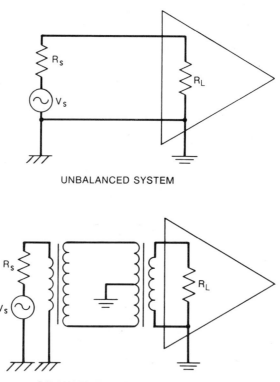

Figure 4-6. *Transmission line portion of a circuit can be balanced by using two transformers.*

*For measuring the balance of individual circuits or components the procedures of IEEE Standard 455-1976 should be used.

One way to guarantee good overall system balance is to specify the balance for each component considerably higher than the desired system balance. This method, however, may not produce the most economical design.

In an otherwise unbalanced system, the transmission line portion of the circuit can be balanced by using two transformers, as shown in Fig. 4-6. Since the conductors are usually the most susceptible to noise pickup, this system can be very useful in reducing noise. The transformers also break any ground loops and, therefore, eliminate the noise due to a difference in ground potential between the load and source.

POWER SUPPLY DECOUPLING

In most electronic systems, the dc power supply and distribution systems are common to many other circuits. It is very important, therefore, to design the dc power system so that it is not a channel for noise coupling between the circuits connected to it. The object of a power distribution system is to supply a nearly constant dc voltage to all loads under conditions of varying load currents. In addition, any ac signals generated by the load should not generate an ac voltage across the dc power bus.

Ideally, a power supply is a zero impedance source of voltage. Unfortunately, practical supplies do not have zero impedance, so they represent a source of coupling between the circuits using them. Not only do the supplies have finite impedance, but the conductors used to connect them to the circuit add to this impedance. Figure 4-7 shows a typical power distribution system as it might appear on a schematic. The dc source—a battery, power supply, or converter—is fused and connected to the variable load R_L by a pair of conductors. A local bypass capacitor C may also be connected across the load.

For detailed analysis the simplified circuit of Fig. 4-7 must be expanded into the circuit of Fig. 4-8. Here, R_s represents the source impedance of the power supply and is a function of the power supply regulation. Resistor R_F represents the resistance of the fuse. Components R_T, L_T, and C_T represent the distributed resistance, inductance, and capacitance; respectively, of the transmission line used to connect the power source to the load. Generator

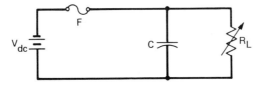

Figure 4-7. *DC power distribution system, as shown on a schematic.*

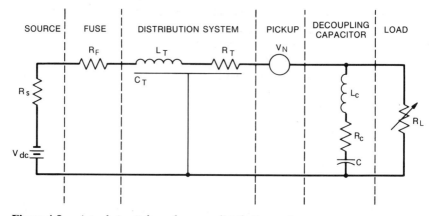

Figure 4-8. *Actual circuit for a dc power distribution system.*

V_N is a lumped noise voltage representing noise coupling into the wiring from other circuits. The bypass capacitor C has resistance R_c and inductance L_c associated with it. Resistor R_L represents the load.

The noise pickup V_N can be minimized by the techniques previously covered in Chapters 2 and 3. The effect of the filter capacitor is discussed in a later section. When the filter capacitor and V_N are eliminated from Fig. 4-8, the circuit of Fig. 4-9 remains. This circuit is used to determine the performance of the power distribution system. The problem can be simplified further by dividing the analysis of Fig. 4-9 into two parts. First, determine the static or dc performance of the system, and second, determine the transient or noise performance of the system.

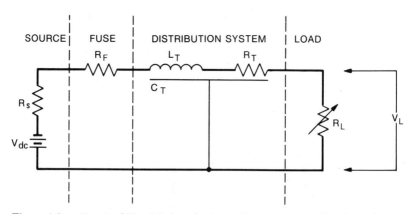

Figure 4-9. *Circuit of Fig. 4-8, less the decoupling capacitor and pickup voltage.*

The static voltage drop is determined by the maximum load current and the resistances R_s, R_F, and R_T. The source resistance R_s can be decreased by improving the regulation of the power supply. The resistance R_T of the power distribution line is a function of the cross sectional area A and length l of the conductors and the resistivity (ρ) of the conductor material,

$$R_T = \rho \frac{l}{A}. \tag{4-7}$$

The resistivity ρ equals 1.724×10^{-6} Ω-cm for copper. The minimum dc load voltage is

$$V_{L(\min)} = V_{dc(\min)} - I_{L(\max)}(R_s + R_F + R_T)_{\max}. \tag{4-8}$$

Transient noise voltages on the power distribution circuit are produced by sudden changes in the current demand of the load. If the current change is assumed to be instantaneous, the magnitude of the resulting voltage change is a function of the characteristic impedance (Z_o) of the transmission line:

$$Z_o = \sqrt{\frac{L_T}{C_T}}. \tag{4-9}$$

The instantaneous voltage change ΔV_L across the load will then be

$$\Delta V_L = \Delta I_L Z_o. \tag{4-10}$$

The assumption of an instantaneous change in current is realistic for digital circuits, but not necessarily so for analog circuits. Even in the case of analog circuits, however, the characteristic impedance of the dc power distribution transmission line can be used as a figure of merit for comparing the noise performance of various power distribution systems. For best noise performance, a power transmission line with as low a characteristic impedance as possible is desired—typically a few ohms or less. Equation 4-9 shows that the line should, therefore, have high capacitance and low inductance.

The inductance can be reduced by using a rectangular cross section conductor instead of a round conductor and by having the two conductors as close together as possible. Both of these efforts also increase the capacitance of the line, as does insulating the conductors with a material having a high dielectric constant. Figure 4-10 gives the characteristic impedance for various conductor configurations. These equations can be used even if the inequalities listed in the figure are not satisfied. Under these conditions, however, the equations give higher values of Z_o than the

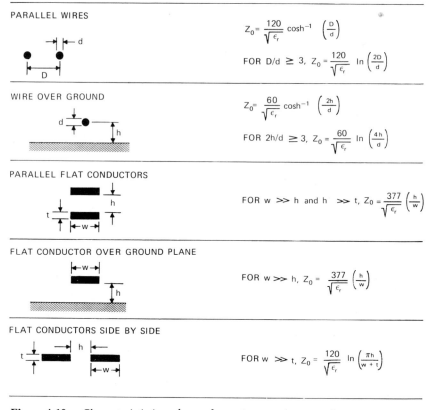

Figure 4-10. *Characteristic impedances for various conductor configurations.*

actual value since they neglect fringing. Values of the relative dielectric constant (ϵ_r) for various materials are listed in Table 4-1. The optimum power distribution line would be one with parallel flat conductors, as wide as possible, placed one on top of each other, and as close together as possible.

To demonstrate the difficulty involved in providing power distribution systems with very low impedance, it is helpful to work some numerical examples. First consider two round parallel wires spaced 1.5 times their diameter apart with Teflon dielectric. The characteristic impedance is as follows:

$$Z_o = \frac{120}{(2.1)^{1/2}} \cosh^{-1}(1.5) = 80\,\Omega.$$

If the dielectric had been air, the impedance would be 115 Ω.

Table 4-1   Relative Dielectric Constants of Various Materials

Material	ϵ_r
Air	1.0
Polyethylene foam	1.6
Cellular polyethylene	1.8
Teflon	2.1
Polyethylene	2.3
Polystyrene	2.5
Nylon	3.0
Silicone rubber	3.1
Polyvinylchloride (PVC)	3.5
Epoxy resin	3.6
Delrin	3.7
Epoxy glass	4.7
Mylar	5.0
Polyurethane	7.0

As a second example, take two flat conductors 0.0027-in. thick by 0.05-in. wide placed side by side on a printed wiring board made of an epoxy resin. If they are spaced 0.05-in. apart, the characteristic impedance is

$$Z_o = \frac{120}{(3.6)^{1/2}} \ln \frac{0.05\pi}{0.0527} = 69 \ \Omega.$$

For an air dielectric the impedance would be 131 Ω. The actual impedance is somewhere between these two values, since on a printed wiring board, part of the field is in air and part is in epoxy.

Both of the above examples are common configurations, and neither one produced a very low impedance transmission line. If, however, two flat conductors 0.25-in. wide are placed one on top of the other and separated by a thin (0.005 in.) sheet of mylar, the characteristic impedance is

$$Z_o = \frac{377}{(5)^{1/2}} \left(\frac{0.005}{0.25} \right) = 3.4 \ \Omega.$$

Such a configuration makes a good low impedance dc power distribution line. Commercial bus bars of this type are available for use with integrated circuits on printed circuit boards, as shown in Fig. 4-11.

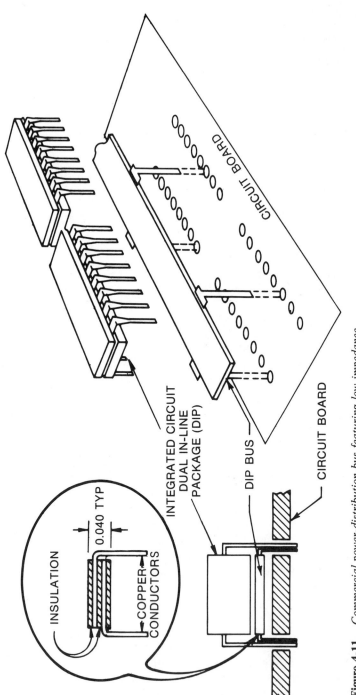

Figure 4-11. *Commercial power-distribution bus featuring low impedance.*

CIRCUIT BOARD

INTEGRATED CIRCUIT
DUAL IN-LINE
PACKAGE (DIP)

DIP BUS

CIRCUIT BOARD

INSULATION

0.040 TYP

COPPER
CONDUCTORS

The difficulty of obtaining a transmission line with sufficiently low impedance usually makes it necessary to place a decoupling capacitor across the power bus at the load to provide a low impedance. Although this is a good practice, a capacitor will not maintain a low impedance at all frequencies because of its series inductance. If the transmission line is designed properly, however, it maintains a low impedance even at high frequencies.

DECOUPLING FILTERS

Since the power supply and its distribution system are not an ideal voltage source it is a good practice to provide some decoupling at each circuit or group of circuits to minimize noise coupling through the supply system. This is especially important when the power supply and its distribution system are not under the control of the designer of the power consuming circuit.

Resistor-capacitor and inductor-capacitor decoupling networks can be used to isolate circuits from the power supply, to eliminate coupling between circuits, and to keep power supply noise from entering the circuit. Neglecting the dashed capacitor, Fig. 4-12 shows two such arrangements. When the R–C filter of Fig. 4-12A is used, the voltage drop in the resistor causes a decrease in power supply voltage. This drop usually limits the amount of filtering possible with this configuration.

The L–C filter of Fig. 4-12B provides more filtering—especially at high frequencies—for the same loss in power supply voltage. The L–C filter, however, has a resonance frequency,

$$f_r = \frac{1}{2\pi\sqrt{LC}} , \qquad (4\text{-}11)$$

at which the signal transmitted through the filter may be greater than if no filter was used. Care must be exercised to see that this resonant frequency is well below the passband of the circuit connected to the filter. The amount of gain in an L–C filter at resonance is inversely proportional to the damping factor

$$\zeta = \frac{R}{2}\sqrt{\frac{C}{L}} , \qquad (4\text{-}12)$$

where R is the resistance of the inductor. The response of an L–C filter near resonance is shown in Fig. 4-13. In order to limit the gain at resonance to less than 2 dB, the damping factor must be greater than 0.5. Additional resistance can be added in series with the inductor, if required,

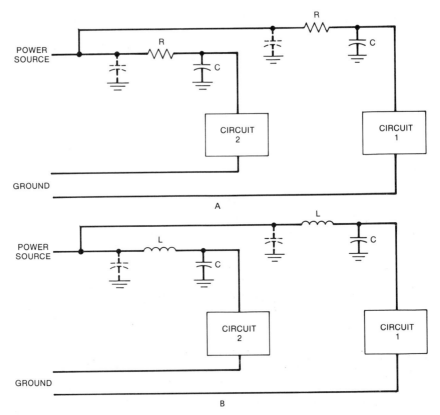

Figure 4-12. *Circuit decoupling with: (A) resistance-capacitance, and (B) inductance-capacitance decoupling networks.*

to increase the damping. The inductor used must also be able to pass the direct current required by the circuit without saturating. A second capacitor, such as those shown dashed in Fig. 4-12, can be added to each section to increase filtering to noise being fed back to the power supply from the circuit. This turns the filter into a pi-network.

When considering noise, a dissipative filter such as the $R–C$ circuit shown in Fig. 4-12A is preferred to a reactive filter, such as the $L–C$ circuit of Fig. 4-12B. In the dissipative filter, the undesirable noise voltage is converted to heat and eliminated as a noise source. In the reactive filter, however, the noise voltage is just moved around. Instead of appearing across the load, the noise voltage now appears across the inductor, where it may be radiated and become a problem in some other part of the circuit. It might then be necessary to shield the inductor to eliminate the radiation.

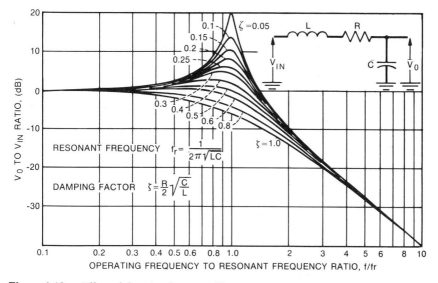

Figure 4-13. *Effect of damping factor on filter response.*

AMPLIFIER DECOUPLING

Even if only a single amplifier is connected to a power supply, consideration of the impedance of the power supply is usually required. Figure 4-14 shows a schematic of a typical two-stage transistor amplifier. When this circuit is analyzed, it is assumed that the ac impedance between the power supply lead and ground is zero. This is hard to guarantee (because the power supply and its wiring has inductance and resistance) unless a decoupling capacitor is placed between the power supply and ground at the amplifier. This capacitor should serve as a short circuit through the frequency range over which the amplifier is capable of producing gain. This frequency range may be much wider than that of the signal being amplified. If this short circuit is not provided across the power supply terminals of the amplifier, the circuit can produce an ac voltage gain to the power supply lead. This signal voltage on the power supply lead can then be fed back to the amplifier input through resistor R_{b1} and possibly cause oscillation.

An emitter follower, feeding a capacitive load such as a transmission line, is especially susceptible to high frequency oscillation due to inadequate power supply decoupling.* Figure 4-15 shows such a circuit. The collector impedance Z_c, consisting of the parasitic inductance of the power

---

*Even with a zero impedance power supply, an emitter follower with a capacitive load can oscillate if improperly designed. See Joyce and Clarke (1961) pp. 264–269.

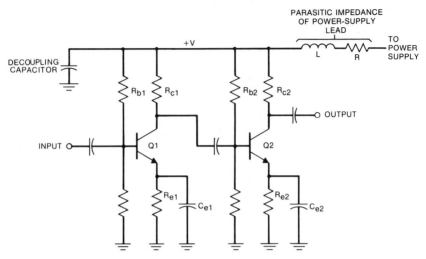

Figure 4-14. *Power supply decoupling for two-stage transistor amplifier.*

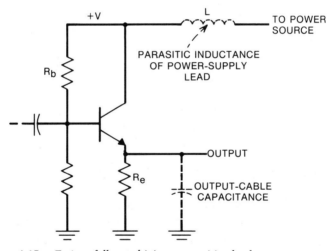

Figure 4-15. *Emitter follower driving a capacitive load.*

supply leads increases with frequency, and the emitter impedance Z_e decreases with frequency due to the cable capacitance. At high frequency, the transistor, therefore, has a large voltage gain to its collector,

$$\text{Voltage Gain} \approx \frac{Z_c}{Z_e}. \tag{4-13}$$

This provides ac feedback around the transistor through the base resistance R_b, thus allowing the possibility of oscillation. If previous stages of the same amplifier are connected to the same power line, the feedback can come through the preceding stages and the possibility of oscillation is greater. The oscillation is often a function of the presence or absence of the output cable, since the cable affects the emitter capacitance and hence the high frequency gain and phase shift through the transistor.

To eliminate the effect of the parasitic lead inductance, a good high frequency ground must be placed at the power terminal of the amplifier. This can be accomplished by connecting a capacitor between the power lead and a good high frequency ground at the amplifier, as shown in Fig. 4-16. The value of this capacitor should be considerably greater than the maximum value of the emitter capacitance C_1. This guarantees that the high frequency gain to the collector of the transistor is always less than one.

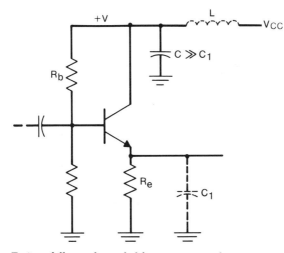

Figure 4-16. *Emitter follower decoupled from power supply.*

Even placing a capacitor across the amplifier power supply terminals cannot guarantee zero impedance. Therefore, some signal will always be fed back to the input circuit over the power supply lead. In amplifiers with gains less than 60 dB, this feedback is usually not enough to cause oscillation. In higher gain amplifiers, this feedback from output to input through the power supply can often cause oscillation. The feedback can be eliminated with an $R-C$ filter in the power supply to the first stage, as shown in Fig. 4-17. The dc voltage drop across the resistor is not detrimental since the first stage operates at a low signal level and therefore, does not require as much dc supply voltage.

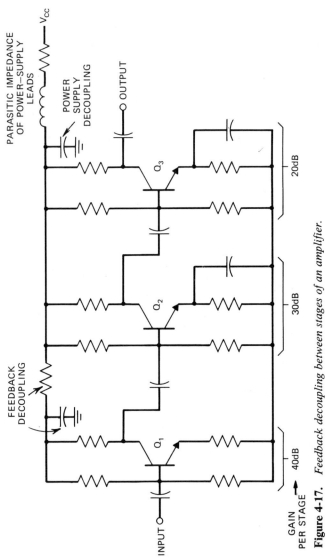

Figure 4-17. *Feedback decoupling between stages of an amplifier.*

HIGH FREQUENCY FILTERING

Metallic enclosures are often used as shields to prevent noisy, or high frequency, circuits from radiating noise. For these shields to be effective, all leads entering or leaving the shielded enclosure should be filtered to prevent them from conducting noise out of the shield. At audio frequencies, normal decoupling filters such as those previously described for power supplies are satisfactory. However, at high frequencies, special care must be taken to guarantee the effectiveness of the filter. Feed-through capacitors* should be used where the conductor passes through the shield, and a mica or ceramic capacitor, with short leads, should be connected between the conductor and ground at the circuit end. This connection, plus three other ways to filter a power supply lead to a high frequency circuit are shown in Fig. 4-18. Shielding the conductor inside the enclosure decreases the amount of noise picked up by the conductor. Additional filtering can be obtained by using a $C–L–C$ pi-filter with two capacitors and an inductor (rf choke). This pi-filter can be further improved by enclosing the choke in a separate shield, inside the primary shield, to prevent it from picking up noise. In all the above filters, the lead lengths on the capacitors and shield grounds must be kept as short as possible.

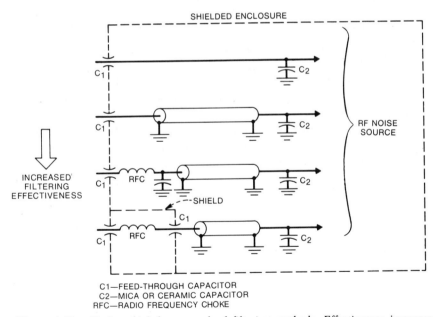

C1—FEED-THROUGH CAPACITOR
C2—MICA OR CERAMIC CAPACITOR
RFC—RADIO FREQUENCY CHOKE

Figure 4-18. *Various high-frequency lead filtering methods. Effectiveness increases from top toward bottom.*

*See Chapter 5 for a discussion of feed-through capacitors.

SYSTEM BANDWIDTH

One simple but often overlooked method of minimizing noise in a system is to limit the system bandwidth to that required by the signal. Use of a circuit bandwidth greater than that required by the signal allows additional noise frequencies to enter the circuit. The same principle applies in the case of digital logic circuits. High speed logic is much more likely to generate high frequency interference than its lower speed counterpart.

MODULATION AND CODING

The susceptibility of a system to interference is a function not only of the shielding, grounding, and so on, but also of the coding or modulating scheme used for the signal. Modulation systems such as amplitude, frequency, and phase all have inherent noise immunity. For example, frequency modulation is very insensitive to amplitude noise disturbances. Digital techniques such as pulse amplitude, pulse width, and pulse repetition frequency coding may be used to increase noise immunity. The noise advantages of various coding and modulation schemes are adequately covered in the literature [Panter (1965); Schwartz (1970); and Schwartz et al. (1966)] and are not repeated here.

DIGITAL CIRCUITS

Although all the previously discussed noise reduction techniques apply to digital as well as analog circuits it is useful to see how some peculiarities of digital circuits affect circuit noise performance. Small integrated circuit logic gates, which draw only a few milliamperes of direct current, do not seem, at first glance, to be serious noise sources. However, when their high switching speed is considered, a problem becomes obvious. For example, a typical TTL (transistor–transistor logic) gate may draw 5 mA from the dc supply in the ON state and 1 mA in the OFF state. This is a current change of only 4 mA, but it may take place in 5 ns. If the power supply wiring has an inductance of 0.5 μH, the noise voltage generated across the power supply wiring when this one gate changes state is

$$V = L\frac{di}{dt} = 0.5 \times 10^{-6}\frac{4 \times 10^{-3}}{5 \times 10^{-9}} = 0.4 \text{ V}. \qquad (4\text{-}14)$$

Multiplying this by the large number of gates in a typical system and realizing that the typical supply voltage for such a logic circuit is only 5 V, it becomes obvious that there can be a major noise problem.

When the designer deals with pulses and digital circuits, it is useful to have an approximate idea of the frequency spectrum contained in the pulses. If pulse rise time t_r is known, an approximate equation can be used to relate it to an equivalent maximum frequency:

$$f_{max} \approx \frac{1}{2\pi t_r} . \qquad (4\text{-}15)$$

As an example, a switching time of 5 ns is equivalent to a maximum frequency of 31.8 MHz. The actual frequency spectrum of a pulse depends on the pulse shape. For nonrepetitive pulses, all frequencies from dc up to f_{max} are present. For repetitive pulses, all frequencies from the repetition frequency up to f_{max} are present.

Figure 4–19 is a basic schematic drawing of a TTL gate. When one of the inputs is grounded, transistor $Q1$ turns on, which turns transistors $Q2$ and $Q4$ off. Transistor $Q3$ is then driven on by the current through $R2$. Transistor $Q3$ amplifies this current and provides the charging current for the load C_L, limited only by resistor $R4$. For fast switching, $R4$ must be a small resistor (50–500 ohms typically).

The use of output transistors $Q3$ and $Q4$ connected as shown in Figure 4–19 is called a totem pole output. This totem pole causes one of the primary noise problems associated with TTL logic devices. When the output is in the "1" state, transistor $Q3$ is on and $Q4$ is off. Conversely, when the output is in the "0" state, transistor $Q3$ is off and $Q4$ is on. Both of these states provide a high impedance between V_{cc} and ground. However, when the gate is switching from one state to the other, there is a short time during which both transistors $Q3$ and $Q4$ are on. This results in a low-impedance connection between the power supply V_{cc} and ground. This

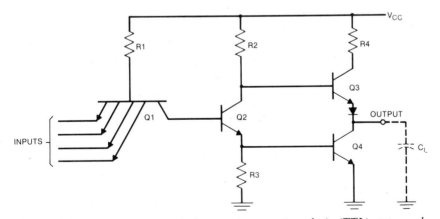

Figure 4-19. *Basic gate schematic for transistor-transistor logic (TTL): totem pole output configuration.*

can result in a power supply current spike of from 10 to 100 mA each time the circuit changes state. If an additional source of current, such as a capacitor, is not located near the gate, disruptive power supply voltage transients will occur. This applies not only to TTL circuits but to any other type which use the totem pole output circuit.

Because of these current spikes and large charging currents required by capacitive loads, *high-frequency decoupling capacitors should be used at each integrated circuit package.* Typically, 0.01–0.02 μF should be used per package; they should be located next to the packages and not lumped together at one point on the board. If several packages are located close to each other on the board, however, one capacitor may be used for up to five packages. An additional power supply decoupling capacitor of 10–100 μF should also be used on each printed wiring board where the power comes onto the board. The dc power distribution conductors should be laid out to form the lowest impedance transmission line as possible; this is a case where the bus assembly of Fig. 4–11 may be useful. Interconnecting signal lines between gates should be less than about 6 in. to avoid ringing.

Due to the high frequencies produced by very fast logic, these circuits should be treated as high frequency circuits. *A good ground is necessary on any printed circuit board containing a large number of logic circuits.* The ground can be either a low-impedance ground bus or a ground plane covering 60 percent or more of the board area. The ground plane provides a low inductance ground return for the power supply and allows for the possibility of using constant impedance lines for signal interconnections. Power buses used for this type logic should be as wide as possible (0.1 in. or greater) in order to minimize their inductance. Ground loops on or off the printed circuit board should be avoided.

To prevent unnecessary switching and noise generation, all unused inputs should be connected to a reference, not left open. The unused inputs are normally connected to the $B+$ voltage, usually through a series resistor, or connected to ground.

Table 4–2 lists typical characteristics of various digital integrated circuit logic families. The noise generation column rates the various families on the basis of radiated and conducted noise. TTL generates the most noise, and HTL generates the least. The speed of the gate is proportional to the propagation delay, listed in Table 4–2. ECL is the fastest and HTL is the slowest. In general, the faster the switching speed, the more noise there will be. This is not true in the case of ECL circuits, however. Due to their balanced configuration the supply current to the gate is the same in both ON and OFF states. There is a zero rate of change of current as the gate switches and, therefore, low noise generation. The commonly used TTL is slightly slower than ECL, but it generates about 10 times as much noise.

Table 4–2   Typical Characteristics of Various Digital Logic Families

Logic families	Propagation delay (ns)	Supply Voltage V_{cc} (V)	DC Noise margin (V)	Noise generation
Emitter Coupled Logic (ECL)	3	5	0.2	Low-medium
Transistor-Transistor Logic (TTL)	10	5	1.0	High
Resistor-Transistor Logic (RTL)	25	4	0.5	Medium
Diode-Transistor Logic (DTL)	30	5	1.0	Medium
Complementary Metal Oxide (CMOS)	35	16	7.0	Medium-high
High Threshold Logic (HTL)	85	15	6.0	Low

Another important characteristic of digital logic circuitry is its susceptibility to noise signals. The dc noise margin as listed in Table 4–2 is defined as the magnitude of pulse noise voltage, which when appearing at the input of a gate and riding on the worst-case logic level, will cause the gate to trigger. The dc noise margin is also applicable to any noise pulses whose width is greater than the propagation delay of the logic. Noise margin to noise spikes with pulse widths less than the propagation delay will be greater than the dc margins listed. RTL is the most susceptible to noise and HTL and CMOS logic are the least susceptible.

SUMMARY

- In a balanced system both resistive and reactive balance must be maintained.
- The greater the degree of balance, the less noise that will couple into the system.
- Balancing can be used with shielding to provide additional noise reduction.
- The lower the characteristic impedance of a dc power distribution circuit, the less the noise coupling over it.
- Since most dc power buses do not provide a low impedance, a decoupling capacitor should be used at each load.
- From a noise point of view, a dissipative filter is preferred to a reactive filter.
- The bandwidth of a system should be limited to that required to transmit the signal in order to minimize noise.

- High speed digital logic can be a source of magnetic noise fields due to its high switching speed.
- Digital logic having a totem pole output circuit produces a low impedance across the power supply during switching.
- High frequency decoupling capacitors should be used across the power line at each IC digital logic package.
- A good ground is necessary on a printed circuit board containing a large number of digital logic packages.

BIBLIOGRAPHY

Alfke, P., and Larsen, I., *The TTL Application Handbook*, Chapter 15 (TTL Characteristics) Fairchild Semiconductor, Mountain View, California August, 1973.

Allan, A., "Noise Immunity Comparison of CMOS Versus Popular Bipolar Logic Families," Motorola Application Note AN-707, 1973.

Balph, T., "Avoid ECL 10,000 Wiring Problems," *Electronic Design*, September, 1972.

Balph, T., "Use ECL 10,000 Layout Rules," *Electronic Design*, August, 1972.

Boaen, V., "Designing Logic Circuits for High Noise Immunity," *IEEE Spectrum*, January, 1973.

Buchman, A. S. "Noise Control in Low Level Data Systems," *Electromechanical Design*, September, 1962.

Costa, D. P., "RFI Suppression, Part II," *Electromechanical Design*, Vol. 11, December, 1967.

Ficchi, R. O., *Practical Design for Electromagnetic Compatibility*, Hayden Book Co., New York, 1971.

Guzik, S. W., and McLellan, D. W., "A Proposed Standard Method of Measurement of Longitudinal Balance of Telephone System Components," National Telecommunications Conference, San Diego, California, 1974.

Joyce, M. V. and Clarke, K. K., *Transistor Circuit Analysis*, Addison-Wesley, Reading, Massachusettes, 1961.

Maul, L., "ECL 10,000 Layout and Loading Rules,"*EDN*, August, 1973.

Maul, L., "Use ECL for Your High-Speed Designs," *EDN*, July 20, 1973.

Nalle, D., "Elimination of Noise in Low Level Circuits," *ISA Journal*, Vol. 12, August, 1965.

Panter, P. F., *Modulation, Noise, and Spectral Analysis*, McGraw-Hill, New York, 1965.

Schwartz, M., *Information Transmission, Modulation, and Noise*, Second Edition, McGraw-Hill, New York, 1970.

Schwartz, M., Bennett, W. R., and Stein, S., *Communication Systems and Techniques*, McGraw-Hill, New York, 1966.

White, D. R. J., *Electromagnetic Interference and Compatibility*, Vol. 3 (EMI Control Methods and Techniques), Don White Consultants, Germantown, Maryland, 1973.

5 PASSIVE COMPONENTS

Since actual components are not "ideal," their characteristics deviate from those of the theoretical components. Understanding these deviations is important in determining the proper application of various components. This chapter is devoted to those characteristics of passive electronic components which affect their noise performance or their use in noise reduction circuitry.

CAPACITORS

Capacitors are most frequently categorized by the dielectric material from which they are made. Different types of capacitors have characteristics that make them suitable for certain applications but not for others. An actual capacitor is not a pure capacitance, but it also has both resistance and inductance, as shown in the equivalent circuit in Fig. 5–1. The inductance L is due to leads as well as the capacitor structure. Resistance R_2 is the parallel leakage and a function of the volume resistivity of the dielectric material. R_1 is the effective series resistance of the capacitor and a function of the dissipation factor of the capacitor.

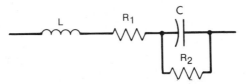

Figure 5-1. *Equivalent circuit for a capacitor.*

Operating frequency is one of the most important considerations in choosing a capacitor type. The maximum effective frequency for a capacitor is usually limited by the inductance of the capacitor and its leads. At some frequency, the capacitor becomes self resonant with its inductance. At frequencies above self resonance, the capacitor has inductive reactance and an impedance increasing with frequency. Figure 5–2 shows how the impedance of a 0.1 μF paper capacitor changes with frequency. As can be seen, this capacitor is self resonant at about 2.5 MHz.

115

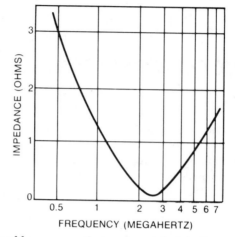

Figure 5-2. *Effect of frequency on the impedance of a 0.1 µF paper capacitor.*

Figure 5–3 shows the approximate usable frequency ranges for various types of capacitors. The high frequency limit is due to self resonance or an increase in the dissipation factor at high frequencies. The low frequency limit is determined by the largest practical capacitance value available.

The primary advantage of an electrolytic capacitor is the large capacitance value that can be put in a small case. The capacitance-to-volume ratio is larger for an electrolytic than for any other capacitor type. An aluminum electrolytic capacitor, however, may have as much as one ohm series resistance. Typical values are about 0.1 Ω. The series resistance increases with increasing frequency—due to dielectric losses—and with decreasing temperature. At −40°C, the series resistance may be 10–100 times the value at 25°C. Due to their large size, aluminum electrolytics also have a large inductance. They are, therefore, low-frequency capacitors and should normally not be used at frequencies above 25 kHz. They are most often used for low frequency filtering, bypassing, and coupling. For use at higher frequencies, they should be bypassed by a low value, low inductance, capacitor.

One disadvantage of electrolytic capacitors is the fact that they are polarized, and a dc voltage of the proper polarity must be maintained across them. For maximum life they should be operated at no greater than 80% of their rated voltage. Operating at less than 80% of their rated voltage does not provide any additional reliability. A nonpolarized capacitor can be made by connecting two equal value electrolytics in series, but poled in opposite directions. The resulting capacitance is one half that of the individual capacitors, and the voltage rating is equal to that of one of the individual capacitors.

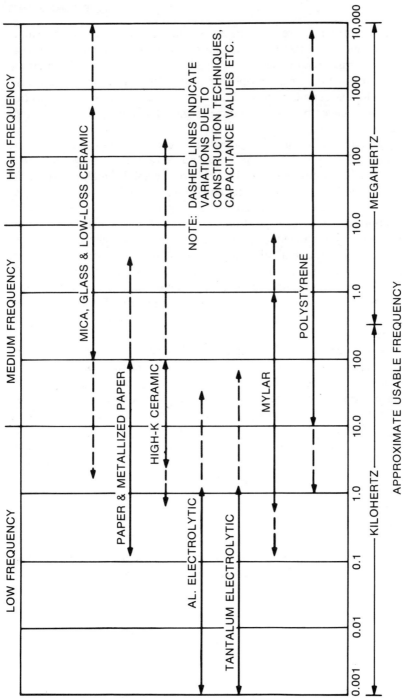

Figure 5-3. *Approximate usable frequency ranges for various types of capacitors.*

117

When electrolytics are used in ac or pulsating dc circuits, the ripple voltage should not exceed the maximum rated ripple voltage; otherwise, excessive internal heating occurs. Normally, the maximum ripple voltage is specified at 120 Hz, typical of operation as a filter capacitor in a full-wave rectifier circuit. Temperature is the primary cause of aging, and electrolytics should never be operated outside their recommended temperature ratings.

Dry tantalum electrolytic capitors have characteristics similar to the aluminum electrolytics. They have less series resistance, however, and a higher capacitance-to-volume ratio. Some of the solid tantalum capacitors have reasonably low inductance and can be used at slightly higher frequencies than aluminum electrolytics. In general, they are more stable than aluminum with respect to time, temperature, and shock.

Paper and mylar capacitors have series resistances considerably less than that of electrolytics, but still have moderately high inductance. Their capacitance-to-volume ratio is less than that of electrolytics, and they are usually available in values up to a few microfarads. They are medium frequency capacitors useful up to a few megahertz. Paper and mylar capacitors are typically used for filtering, bypassing, coupling, timing, and noise suppression.

Tubular capacitors such as paper or mylar usually have a band around one end, as shown in Fig. 5–4. The lead connected to the banded end is connected to the outside foil of the capacitor. The banded end should be connected to ground, or to a common reference potential whenever possible. In this way, the outside foil of the capacitor can act as a shield to minimize electric-field coupling from the capacitor.

Mica and ceramic capacitors have very low series resistance and inductance. They are high-frequency capacitors and useful up to about 500 MHz —provided the leads are kept short. These capacitors are normally used for high-frequency filtering, bypassing, coupling, timing, and frequency discrimination. They are normally very stable with respect to time, temperature, and voltage.

High K ceramic capacitors however, are only medium frequency capacitors. They are relatively unstable with respect to time, temperature, and frequency. Their primary advantage is a higher capacitance-to-volume

◄--- BAND

Figure 5-4. *Band on tubular capacitor indicates the end which is connected to the outside foil. This end should be connected to ground.*

ratio, compared to that of standard ceramic capacitors. They are normally used for bypassing, coupling, and blocking. One disadvantage is that they may be damaged by voltage transients. It is, therefore, not recommended that they be used as bypass capacitors directly across a low-impedance power supply.

Polystyrene capacitors have extremely low series resistance and have very stable capacitance-frequency characteristics. They are the closest to the ideal capacitor of all the types listed. Typical applications include filtering, bypassing, coupling, timing, and noise suppression.

No one type of capacitor will provide satisfactory filtering over the entire range from low audio frequencies up to high radio frequencies. In order to provide filtering over this range of frequencies, two different capacitors connected in parallel are normally required. An electrolytic can be used to provide the large capacitance necessary for low frequency filtering. This can then be paralleled with a small value, low inductance mica or ceramic capacitor to provide the low impedance at high frequencies.

The effect of lead length and capacitance value on the self-resonant frequency of a small ceramic capacitor can be seen from Table 5–1. It is clear that at high frequencies, the smallest value capacitor that will do the job is preferable because of its higher self-resonant frequency. The self-resonant frequency can be increased by using a feed-through capacitor designed to mount through, or on, a metal chassis. Figure 5–5 shows such a capacitor mounted in a chassis or shield along with its usual schematic representation. The capacitance is between the leads and the case of the capacitor. Several types of feed-through capacitors are pictured in Fig. 5–6. Feed-through capacitors have very low self inductance, and therefore, can be used at higher frequencies. Figure 5–7 shows the lower impedance obtained by using a feed-through capacitor as compared to a standard capacitor.

Table 5–1   Self-Resonant Frequencies of Ceramic Capacitors

Capacitance value (pf)	Self-resonant frequency (MHz)	
	$\frac{1}{4}$-in. Leads	$\frac{1}{2}$-in. Leads
10,000	12	—
1,000	35	32
500	70	65
100	150	120
50	220	200
10	500	350

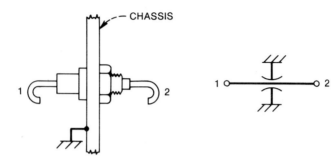

MOUNTED IN CHASSIS   SCHEMATIC REPRESENTATION

Figure 5-5. *Typical feed-through capacitor.*

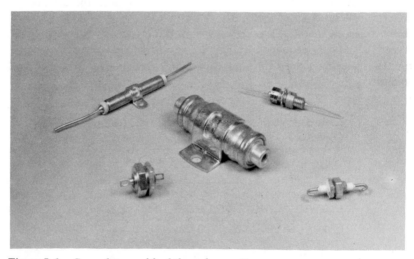

Figure 5-6. *Several types of feed-through capacitors.*

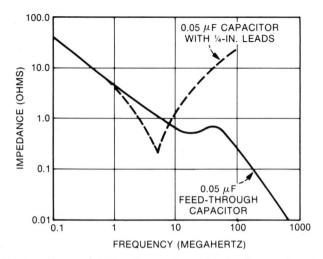

Figure 5-7. *Impedance of 0.05 μF capacitors, showing improved performance of feed-through capacitor.*

INDUCTORS

Inductors may be categorized by the type of core on which they are wound. The two most general categories are air core (any nonmagnetic material fits into this group) and magnetic core. Magnetic core inductors can be further subdivided depending on whether the core is open or closed. An ideal inductor would have only inductance, but an actual inductor also has series resistance, in the wire used to wind it, and distributed capacitance between the windings. This is shown in the equivalent circuit in Fig. 5–8. The capacitance is represented as a lumped shunt capacitor, so there is parallel resonance at some frequency. This resulting resonance point determines the highest frequency at which an inductor can be used.

Another important characteristic of inductors is their susceptibility to, and generation of, stray magnetic fields. *Air core or open magnetic core inductors are most likely to cause interference*, since their flux extends a considerable distance from the inductor, as shown in Fig. 5–9A. Inductors wound on a closed magnetic core have much reduced external magnetic

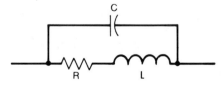
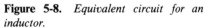

Figure 5-8. *Equivalent circuit for an inductor.*

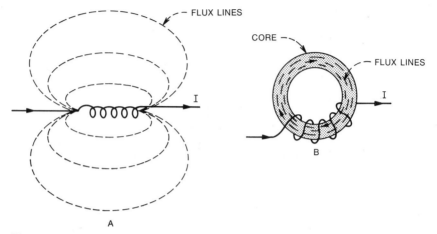

Figure 5-9. *Magnetic fields from: (A) air core, and (B) closed magnetic core inductors.*

fields, since nearly all the flux remains inside the magnetic core, as shown in Fig. 5–9B.

As far as susceptibility to magnetic fields is concerned the magnetic core is more susceptible than the air core inductor. An open magnetic core inductor is the most susceptible, since the core—a low reluctance path—concentrates the external magnetic field and causes more of the flux to flow through the coil. A closed magnetic core is less susceptible than an open core, but more susceptible than an air core.

It is often necessary to shield inductors to confine their magnetic and electric fields within a limited space. Shields made of low resistance material such as copper or aluminum confine the electric fields. At high frequencies these shields also prevent magnetic flux passage, because of the eddy currents set up within the shield. At low frequencies, however, high-permeability magnetic material must be used to confine the magnetic field.*

TRANSFORMERS

Two inductors intentionally coupled together, usually on a magnetic core, form a transformer. Transformers are often used to provide isolation between two circuits. An example is the isolation transformer used to break a ground loop, as shown in Fig. 3–19. In these cases, the only desirable coupling is that which results from the magnetic field. Actual

*See Chapter 6 for a detailed analysis of magnetic-field shielding.

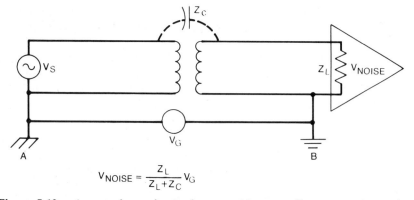

Figure 5-10. *An actual transformer has capacitive as well as magnetic coupling between primary and secondary windings.*

transformers, not being ideal, have capacitance between the primary and secondary windings, as shown in Fig. 5–10, and this allows noise coupling through the transformer.

This coupling can be eliminated by providing an electrostatic, or Faraday, shield (a grounded conductor between the two windings), as shown in Fig. 5–11. If properly designed, this shield does not affect the magnetic coupling, but it eliminates the capacitive coupling provided the shield is grounded. The shield must be grounded at point B in Fig. 5-11. If it is grounded to point A, the shield is at a potential of V_G and still couples

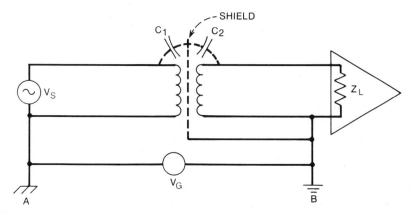

Figure 5-11. *Grounded electrostatic shield between transformer windings breaks the capacitive coupling.*

noise through the capacitor C_2 to the load. Therefore, the transformer should be located near the load in order to simplify the connection between the shield and point B.

Electrostatic shielding may also be obtained with two unshielded transformers, as shown in Fig. 5–12. The primary circuit of T_2 must be grounded, preferably with a center tap. The secondary of T_1, if it has a center tap, may also be grounded to hold one end of C_2 near ground potential. If the transformers do not have center taps, one of the conductors between the transformers can be grounded, as indicated in Fig. 5–12. This configuration is less effective than a transformer with a properly designed electrostatic shield. The configuration of Fig. 5–12 is, however, useful in the laboratory to determine whether an electrostatically shielded transformer can effectively decrease the noise coupling in a circuit.

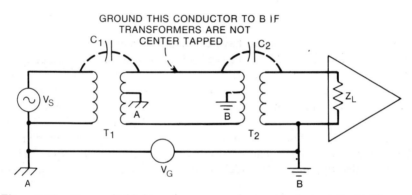

Figure 5-12. *Two unshielded transformers can also provide electrostatic shielding.*

RESISTORS

Fixed resistors can be grouped into three basic classes: (1) wirewound, (2) film type, and (3) composition. The exact equivalent circuit for a resistor depends upon the type of resistor and the manufacturing processes. The circuit of Fig. 5–13 however, is satisfactory in most cases. In a typical composition resistor, the shunt capacitance is in the order of 0.1–0.5 pF. The inductance is primarily lead inductance, except in the case of wirewound resistors, where the resistor body is the largest contributor. Film resistors, due to their spiral or meandering-line construction, have more inductance than carbon resistors. Except for wirewound resistors, or very low value resistors of other types, the inductance can normally be neglected during circuit analysis. The inductance of a resistor does, however, make it susceptible to pick-up from external magnetic fields. Inductance of the external lead can be approximated by using the data in Table 5–3 on page 128.

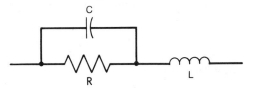

Figure 5-13. *Equivalent circuit for a resistor.*

The shunt capacitance can be important when high value resistors are used. For example, consider a 22-MΩ resistor with 0.5 pF of shunt capacitance. At 145 kHz, the capacitive reactance will be 10% of the resistance. If this resistor is used above this frequency, the capacitance may affect the circuit performance.

Table 5–2 shows measured impedance, magnitude and phase angle, for a $\frac{1}{2}$-W carbon resistor at various frequencies. The nominal resistance value is 1 megohm. Note that at 500 kHz, the impedance has dropped to 560 kΩ and the phase angle has become $-34°$. Capacitive reactance has thus become significant.

Table 5–2   Impedance of a 1-MΩ, $\frac{1}{2}$ W Carbon
Resistor Measured at Various Frequencies

Frequency (kHz)	Impedance	
	Magnitude (kΩ)	Phase Angle (degrees)
1	1000	0
9	1000	-3
10	990	-3
50	920	-11
100	860	-16
200	750	-23
300	670	-28
400	610	-32
500	560	-34

NOISE IN RESISTORS

All resistors, regardless of their construction, generate a noise voltage. This voltage results from thermal noise and other noise sources, such as shot and contact noise. Thermal noise can never be avoided, but the other sources can be minimized or eliminated. The total noise voltage, therefore, is equal to or greater than the thermal noise voltage. This is explained further in Chapter 8.

Of the three basic resistor types, wirewound resistors are the quietest. The noise in a good quality wirewound resistor should be no greater than that due to thermal noise. At the other extreme is the composition resistor, which has the most noise. In addition to thermal noise, composition resistors also have contact noise, since they are made of many individual particles molded together. When no current flows in composition resistors, the noise approaches that of thermal noise. When current flows, additional noise is generated proportional to the current. Figure 5–14 shows the noise generated by a 10 kΩ composition resistor at two current levels. At low frequencies, the noise is predominantly contact noise, which has an inverse frequency characteristic. The frequency at which the noise levels off, at a

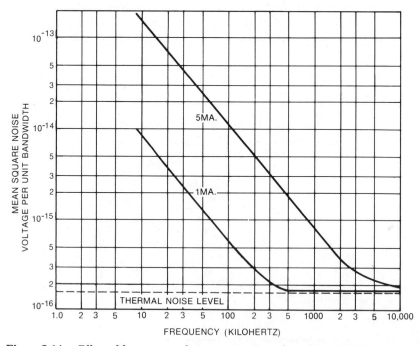

Figure 5-14. *Effect of frequency and current on noise voltage for a 10-kΩ composition resistor.*

value equal to the thermal noise, varies widely between different type resistors and is also dependent on current level.

The noise produced by film-type resistors is much less than that produced by composition resistors, but it is more than that produced by wirewound resistors. The additional noise is again contact noise, but because the material is more homogeneous, the amount of noise is considerably less than for composition resistors.

Another important factor affecting the noise in a resistor is its power rating. If two resistors of the same value and type both dissipate equal power, the resistor with the higher power rating normally has the lower noise. Campbell and Chipman (1949) present data showing approximately a factor of 3 between the rms noise voltage of a $\frac{1}{2}$-W composition resistor versus a 2-W composition resistor operating under the same conditions. This difference is due to the factor K in Eq. 8–19 (Chapter 8), a variable that depends on the geometry of the resistors.

Variable resistors generate all the inherent noises of fixed resistors, but in addition generate noise from wiper contact. This additional noise is directly proportional to current through the resistor and the value of its resistance. To reduce the noise, the current through the resistor and the resistance itself should both be minimized.

CONDUCTORS

Although conductors are not normally considered components, they do have characteristics which are very important to the noise and transient performance of electronic circuits. Inductance is one of the most important of these characteristics. Even at low frequencies, a conductor may have more inductive reactance than resistance.

The external inductance of a straight, round conductor of diameter d, located a distance h above a ground plane is

$$L = \frac{\mu}{2\pi} \ln\left(\frac{4h}{d}\right) \quad H/m. \tag{5-1}$$

This assumes that $h > 1.5\ d$. The permeability of free space (μ) is equal to $4\pi \times 10^{-7}$ H/m. Equation 5–1, therefore, can be rewritten

$$L = 2 \times 10^{-7} \ln\left(\frac{4h}{d}\right) \quad H/m. \tag{5-2}$$

Changing units to microhenries per inch gives

$$L = 0.005 \ln\left(\frac{4h}{d}\right) \quad \mu H/in. \tag{5-3}$$

The above equations represent the external inductance since they do not include the effects of the magnetic field within the conductor itself. The total inductance is actually the sum of the internal plus external inductances. The internal inductance of a straight wire of circular cross section carrying a uniform low frequency current is 1.27×10^{-3} μH/in., independent of wire size. The internal inductance is negligible compared to the external inductance except for very close conductor spacings. The internal inductance is further reduced when higher frequency currents are considered since, due to skin effect, the current is concentrated near the surface of the conductor. The external inductance, therefore, is normally the only inductance of significance.

Table 5–3 lists values of external inductance and resistance for various gauge conductors. The table shows that moving the conductor closer to the ground plane decreases its inductance; this assumes the ground plane is the return circuit. Raising the conductor higher above the ground plane increases the inductance. Beyond a height of a few inches, however, the inductance approaches its free-space value, and increasing the spacing has very little effect on the inductance. This is because almost all the flux produced by current in the conductor is already contained within the loop.

Table 5–3   Inductance and Resistance of Round Conductors

Wire size (AWG)	Diameter (in.)	Resistance (mΩ/in.)	Inductance (μH per in.)		
			0.25 in. above ground plane	0.5 in. above ground plane	1 in. above ground plane
26	0.016	3.39	0.021	0.025	0.028
24	0.020	2.13	0.020	0.023	0.027
22	0.025	1.34	0.019	0.022	0.026
20	0.032	0.85	0.017	0.021	0.024
18	0.040	0.53	0.016	0.020	0.023
14	0.064	0.21	0.014	0.017	0.021
10	0.102	0.08	0.012	0.015	0.019

Table 5–3 also indicates that the larger the conductor the lower the inductance. The inductance and the conductor diameter are logarithmically related. For this reason, low values of inductance are not easily obtained by increasing the conductor diameter. The spacing between conductors affects the external inductance, whereas the cross section affects only the internal inductance. The internal inductance can be reduced by using a flat, rectangular conductor instead of a round one. A hollow round tube also has less inductance than the same size solid conductor.

For two parallel conductors carrying uniform current in opposite directions, the self inductance, neglecting flux in the wires themselves, is

$$L = 0.01 \ln\left(\frac{2D}{d}\right) \quad \mu H/in. \tag{5-4}$$

In Eq. 5–4, D is the center to center spacing and d is the conductor diameter.

Resistance is a second very important characteristic of a conductor. Selection of conductor size is generally determined by the maximum allowable voltage drop in the conductor. The voltage drop is a function of the conductor resistance and the maximum current. Table 5–3 lists the value of dc resistance for various size conductors.

At higher frequencies, resistance of a conductor increases, due to skin effect. Skin effect describes a condition where, due to the magnetic fields produced by current in the conductor, there is a concentration of current near the conductor surface. As the frequency increases, the current is concentrated closer to the surface. This effectively decreases the cross section through which the current flows and, therefore, increases the effective resistance.

For solid, round copper conductors the ac and dc resistances are approximately related by the following expression (ITT, 1968):

$$R_{ac} = \left(0.096d\sqrt{f} + 0.26\right)R_{dc}, \tag{5-5}$$

where d is the conductor diameter in inches and f is the frequency. For $d\sqrt{f}$ greater than ten, Equation 5–5 is accurate within a few percent. For $d\sqrt{f}$ less than ten, the actual ac resistance is greater than that given by Eq. 5–5. If the conductor material is other than copper, the first term of Eq. 5–5 must be multiplied by the factor

$$\sqrt{\frac{\mu_r}{\rho_r}} \ ,$$

where μ_r is the relative permeability of the conductor material and ρ_r is the relative resistivity of the material compared to copper. Due to skin effect, a hollow tube at high frequency has the same ac resistance as a solid conductor.

The ac resistance of a conductor can be decreased by changing the shape of the cross section. A rectangular conductor has inherently lower ac resistance than a round conductor because of its greater surface per unit cross section.

Since a rectangular conductor has less ac resistance and less inductance than a round conductor with the same cross section area, it is a better high-frequency conductor. Flat straps or braid are, therefore, commonly used as ground conductors even in relatively low-frequency circuits.

Equation 5–5 can also be used to determine the approximate ac resistance for any shape conductor by letting

$$d = \frac{(\text{perimeter of cross section})}{\pi} \, . \qquad (5\text{-}6)$$

FERRITE BEADS

Ferrite beads provide an inexpensive and convenient way to add high frequency loss in a circuit without introducing power loss at dc and low frequencies. The beads are small and can be installed simply by slipping them over a component lead or conductor. The beads are most effective in providing attenuation to unwanted signals above 1 MHz. When properly used these beads can provide high frequency decoupling, parasitic suppression, and shielding.

Figure 5–15A shows a small cylindrical bead installed on a conductor and Fig. 5–15B shows the high-frequency equivalent circuit—an inductor in series with a resistor. The resistance comes about due to the high frequency losses in the ferrite material. Figure 5–15C shows typical schematic symbols used for ferrite beads. Most bead manufacturers characterize their components by plotting magnitude of the impedance versus frequency. The magnitude of the impedance is given by

$$|Z| = \sqrt{R^2 + (2\pi f L)^2} \, , \qquad (5\text{-}7)$$

where R is the equivalent resistance of the bead and L is the equivalent

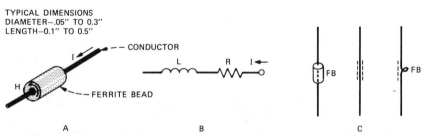

Figure 5-15. *(A) Ferrite bead on conductor; (B) high frequency equivalent circuit; and (C) typical schematic symbols.*

inductance. Figure 5–16 gives data on two typical ferrite beads. Bead number 1 is primarily resistive, whereas bead number 2 is primarily inductive.

Ferrite beads are especially effective when used to damp out high frequency oscillations generated by switching transients or parasitic resonances within a circuit. They are also useful in preventing high frequency noise from being conducted into a circuit on power supply or other leads.

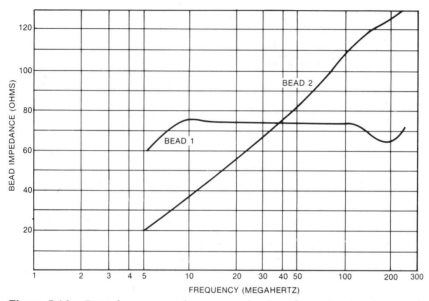

Figure 5-16. *Data for a primarily resistive (bead 1) ferrite bead and primarily inductive (bead 2) ferrite bead.*

Since the impedance of a single bead is limited to about 100 Ω, beads are most effective in low-impedance circuits such as power supplies, class C power amplifiers, resonant circuits, and SCR switching circuits. If a single bead does not provide sufficient attenuation multiple beads may be used. However, if two or three beads do not solve the problem, additional beads are not normally effective.

Figures 5–17 through 5–20 show some typical applications of ferrite beads. In Fig. 5–17, the inductive characteristic of the beads is used to form an *L–C* filter to keep signals from the high frequency oscillator out of the load. A bead with resistive characteristics could also have been used to form a high frequency *R–C* filter without reducing the dc voltage to the load. In Fig. 5–18, a resistive bead is used to damp out the ringing

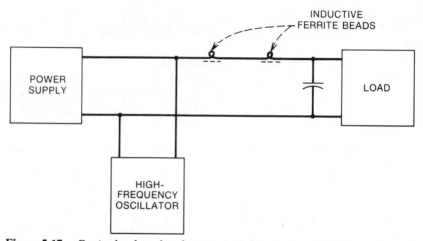

Figure 5-17. *Ferrite bead used to form an L–C filter to keep high frequencies away from load.*

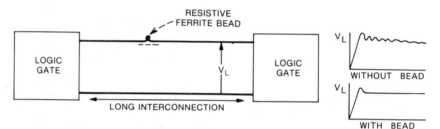

Figure 5-18. *Resistive ferrite bead used to damp out ringing on long line between fast logic gates.*

generated by a long interconnection between two fast logic gates.

Figure 5–19 shows a Class C power amplifier which has an unwanted output signal on a high harmonic frequency due to the parasitic resonant circuit of capacitor C and inductor L. In this case, the inductance of the bead is used to force the harmonic current to flow through the 50-Ω resistor and be dissipated as heat. At the desired operating frequency the impedance of the bead is low, and provides a shunt around the resistor.

Figure 5–20 shows two ferrite beads mounted on a printed circuit board. This circuit is part of the horizontal output for a color television set, and the beads are used to supress parasitic oscillations.

Yet another application for ferrite beads is shown in Fig. 5–21. Figure 5–21A shows a dc servo motor connected to a motor control circuit.

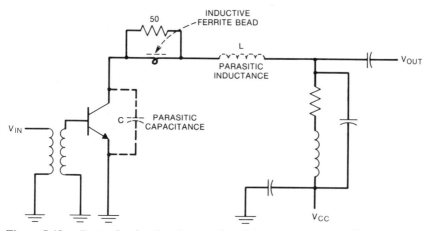

Figure 5-19. *Ferrite bead and resistor used to damp out parasitic oscillation in class C-power output stage.*

Figure 5-20. *Beads installed in color TV set to suppress parasitic oscillations in horizontal output circuit.*

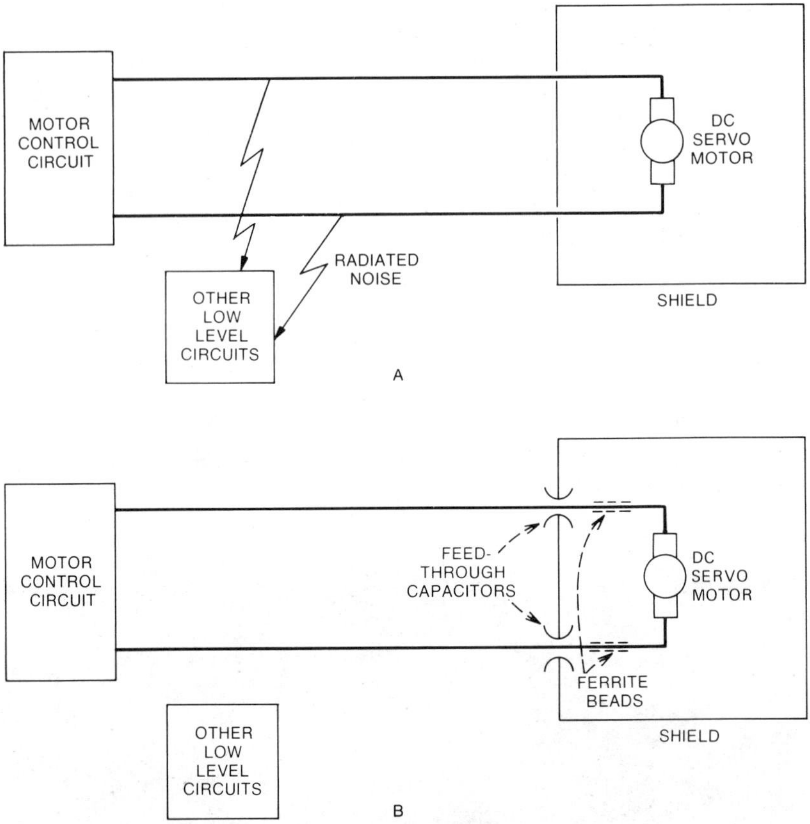

Figure 5-21. *(A) High frequency commutation noise of motor is interfering with low-level circuits; (B) beads used in conjunction with feed-through capacitors to eliminate interference.*

High-frequency commutation noise from the motor is being conducted out of the motor shield on the motor leads, then radiated from the leads to interfere with other low-level circuits. Because of acceleration requirements on the motor, resistance cannot be inserted in the motor leads. The solution in this case was to add two ferrite beads and two feed-through capacitors, as shown in Fig. 5–21B. A photograph of the motor with ferrite beads and feed-through capacitors is also shown in Fig. 5–22.

When using ferrite beads in circuits with dc current, care must be taken to guarantee that the current does not cause saturation of the ferrite material.

Figure 5-22. *Ferrite beads and feed-through capacitors used to reduce motor commutation noise coupling to power leads.*

Since ferrite beads are inductive, they should not be used indiscriminately. In some locations they may do more harm than good; they can themselves produce unwanted resonances in a circuit when misapplied. However, when applied properly they can be a very simple, effective, and inexpensive means to reduce noise and parasitic oscillation.

SUMMARY

- Electrolytics are low frequency capacitors.
- All capacitors become self-resonant at some frequency which limits their high frequency use.

- Mica and ceramic are good high frequency capacitors.
- Air-core inductors generate more noise fields than do closed magnetic-material-core inductors.
- Magnetic core inductors are more susceptible to interfering magnetic fields than are air-core inductors.
- An electrostatic shielded transformer is used to reduce capacitive coupling between the windings.
- All resistor types have the same amount of thermal noise.
- Variable resistors in low-level circuits should be placed so that no dc flows through them.
- Even at low frequencies a conductor normally has more inductive reactance than resistance.
- A flat rectangular conductor will have less ac resistance and inductance, than a round cross section.
- Ferrite beads can be used to add high frequency loss to a circuit without introducing a dc loss.

BIBLIOGRAPHY

Campbell, R. H., Jr., and Chipman, R. A., "Noise From Current-Carrying Resistors, 20-500 KC," *Proceedings of the IRE*, Vol. 37, pp. 938–942, August, 1949.

Costa, D. P., "RFI Suppression, Part II," *Electromechanical Design*, Vol. 11, December, 1967.

Cowdell, R. B., "Don't Experiment With Ferrite Beads," *Electronic Design*, Vol. 17, June 7, 1969.

Dummer, G. W. A., and Nordenberg, H. M., *Fixed and Variable Capacitors*, McGraw-Hill, New York, 1960.

Henney, K., and Walsh, C., *Electronic Components Handbook*, Vol. 1, McGraw-Hill, New York, 1957.

ITT, *Reference Data for Radio Engineers*, Fifth Edition, Howard W. Sams, New York, p. 6-8, 1968.

Rostek, P. M., "Avoid Wiring-Inductance Problems," *Electronic Design*, Vol. 22, December 6, 1974.

Skilling, H. H., *Electric Transmission Lines*, McGraw-Hill, New York, 1951.

6 SHIELDING EFFECTIVENESS OF METALLIC SHEETS

A shield is a metallic partition placed between two regions of space. It is used to control the propagation of electric and magnetic fields from one of the regions to the other. Shields may be used to contain electromagnetic fields if the shield surrounds the noise source as shown in Fig. 6-1. A shield may also be used to keep electromagnetic radiation out of a region, as shown in Fig. 6-2. Circuits, components, cables or complete systems may be shielded, and shielding can be applied to the source, the receiver, or both.

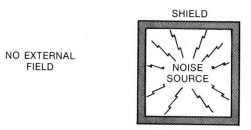

Figure 6-1. *Shield application where a noise source is contained, preventing interference with equipment outside shield.*

NEAR FIELDS AND FAR FIELDS

The characteristics of a field are determined by the source, the media surrounding the source, and the distance between the source and the point of observation. At a point close to the source, the field properties are determined primarily by the source characteristics. Far from the source, the properties of the field depend mainly upon the medium through which the field is propagating. Therefore, the space surrounding a source of radiation can be broken into two regions, as shown in Fig. 6-3. Close to the source is the near, or induction, field. At a distance greater than the wavelength (λ) divided by 2π (approximately one-sixth of a wavelength) is the far, or radiation, field. The region around $\lambda/2\pi$ is the transition region between the near and far fields.

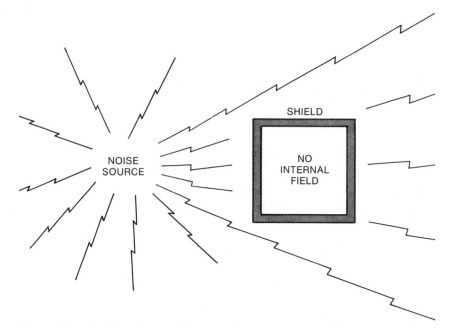

Figure 6-2. *Shield application where interference is prevented by placing a shield around a receiver to prevent noise infiltration.*

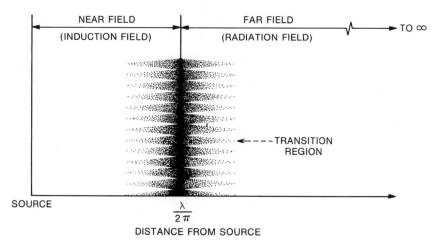

Figure 6-3. *Field character depends upon the distance from the source. The transition from the near to far field occurs at $\lambda/2\pi$.*

The ratio of the electric field (E) to the magnetic field (H) is the wave impedance. In the far field, this ratio E/H equals the characteristic impedance of the medium (e.g., $E/H = Z_0 = 377$ Ω for air or free space). In the near field, the ratio is determined by the characteristics of the source and the distance from the source to where the field is observed. If the source has high current and low voltage ($E/H < 377$) the near field is predominantly magnetic. Conversely, if the source has low current and high voltage ($E/H > 377$) the near field is predominantly electric.

For a rod or straight wire antenna, the source impedance is high. The wave impedance near the antenna—predominantly an electric field—is also high. As distance is increased, the electric field loses some of its intensity as it generates a complementary magnetic field. In the near field, the electric field attenuates at a rate of $(1/r)^3$ whereas the magnetic field attenuates at a rate of $(1/r)^2$. Thus, the wave impedance from a straight wire antenna decreases with distance and asymptotically approaches the impedance of free space in the far field, as shown in Fig. 6-4.

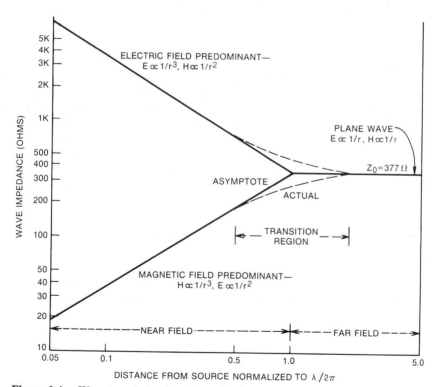

Figure 6-4. *Wave impedance depends on the distance from the source and on whether the field is electric or magnetic.*

For a predominantly magnetic field—such as produced by a loop antenna—the wave impedance near the antenna is low. As the distance from the source increases, the magnetic field attenuates at a rate of $(1/r)^3$ and the electric field attenuates at a rate of $(1/r)^2$. The wave impedance therefore increases with distance and approaches that of free space at a distance of $\lambda/2\pi$. In the far field, both the electric and magnetic fields attenuate at a rate of $1/r$.

At frequencies less than 1 MHz, most coupling within electronic equipment is due to the near field, since the near field at these frequencies extends out to 150 ft or more. At 30 kHz, the near field extends out to 1 mile. Therefore, interference problems within any given equipment should be assumed to be near field problems unless it is clear that they are far field problems.

In the near field, the electric and magnetic fields must be considered separately, since the ratio of the two is not constant. In the far field, however, they combine to form a plane wave having an impedance of 377 Ω. Therefore, when plane waves are discussed, they are assumed to be in the far field. When individual electric and magnetic fields are discussed they are assumed to be in the near field.

SHIELDING EFFECTIVENESS

The following sections discuss the shielding effectiveness of metallic sheets in both the near and far fields. Shielding effectiveness of metallic sheets can be determined by analyzing the problem in either of two ways. One method is to use circuit theory, and the second is to use field theory. In the circuit theory approach, the noise fields induce currents in the shields, and these currents in turn generate additional fields which tend to cancel the original fields in certain regions. An example of this is shown in Fig. 6-5. For most of this chapter, however, we will adopt the more fundamental field theory approach.

Shielding can be specified in terms of the reduction in magnetic and/or electric field strength caused by the shield. It is convenient to express this shielding effectiveness in units of decibels (dB).* Use of dB then permits the shielding produced by various effects or shields to be added to obtain the total shielding. Shielding effectiveness (S) is defined for electric fields as

$$S = 20 \log \frac{E_0}{E_1} \quad \text{dB}, \qquad (6\text{-}1)$$

*See Appendix A for a discussion of the decibel.

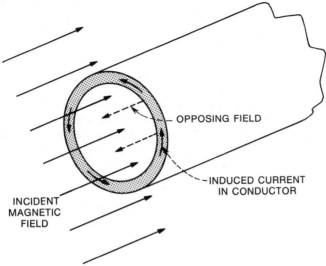

Figure 6-5. *Conducting material can provide magnetic shielding. The incident magnetic field induces currents in the conductor producing an opposing field to cancel the incident field in the region enclosed by the shield.*

and for magnetic fields as

$$S = 20 \log \frac{H_0}{H_1} \quad \text{dB.} \tag{6-2}$$

In the above equations, $E_0(H_0)$ is the incident field strength, and $E_1(H_1)$ is the field strength of the transmitted wave as it emerges from the shield.

Shielding effectiveness varies with frequency, geometry of shield, position within the shield where the field is measured, type of field being attenuated, direction of incidence and polarization. This chapter will consider the shielding provided by a plane sheet of conducting material. This simple geometry serves to introduce general shielding concepts and shows which material properties determine shielding effectiveness, but does not include those effects due to the geometry of the shield. The results of the plane sheet calculations are useful for estimating the relative shielding ability of various materials.

Two types of loss are encountered by an electromagnetic wave striking a metallic surface. The wave is partially reflected from the surface, and the transmitted (nonreflected) portion is attenuated as it passes through the medium. This latter effect, called absorption loss, is the same in either the near or the far field and for electric or magnetic fields. Reflection loss, however, is dependent on the type of field, and the wave impedance.

The total shielding effectiveness of a material is equal to the sum of the absorption loss (A) plus the reflection loss (R) plus a correction factor (B) to account for multiple reflections in thin shields. Total shielding effectiveness, therefore, can be written as

$$S = A + R + B \quad \text{dB.} \qquad (6\text{-}3)$$

All the terms in Eq. 6-3 must be expressed in dB. The multiple reflection factor B can be neglected if the absorption loss A is greater than 10 dB. From a practical point of view, B can also be neglected for electric fields and plane waves.

CHARACTERISTIC AND WAVE IMPEDANCES

The following characteristic constants of a medium are used in this chapter:

Permeability,	$\mu (4\pi \times 10^{-7}$ H/m for free space).
Dielectric constant,	$\epsilon (8.85 \times 10^{-12}$ F/m for free space).
Conductivity,	$\sigma (5.82 \times 10^{7}$ mhos/m for copper).

For any electromagnetic wave, the wave impedance is defined as

$$Z_w = \frac{E}{H} . \qquad (6\text{-}4)$$

The characteristic impedance of a medium is defined (Hayt, 1974) by the following expression:

$$Z_0 = \sqrt{\frac{j\omega\mu}{\sigma + j\omega\epsilon}} . \qquad (6\text{-}5)$$

In the case of a plane wave in the far field, Z_0 is also equal to the wave impedance Z_w. For insulators ($\sigma \ll j\omega\epsilon$) the characteristic impedance becomes

$$Z_0 = \sqrt{\frac{\mu}{\epsilon}} . \qquad (6\text{-}6)$$

In free space Z_0 equals 377 Ω. In the case of conductors ($\sigma \gg j\omega\epsilon$), the characteristic impedance is called the shield impedance Z_s and becomes

$$Z_s = \sqrt{\frac{j\omega\mu}{\sigma}} = \sqrt{\frac{\omega\mu}{2\sigma}} (1+j), \qquad (6\text{-}7a)$$

$$|Z_s| = \sqrt{\frac{\omega\mu}{\sigma}} \ . \tag{6-7b}$$

For copper at 1 kHz, $|Z_s|$ equals 1.16×10^{-5} Ω. Substituting numerical values for the constants of Eq. 6-7b gives the following results.
For copper,

$$|Z_s| = 3.68 \times 10^{-7} \sqrt{f} \ . \tag{6-8a}$$

For aluminum,

$$|Z_s| = 4.71 \times 10^{-7} \sqrt{f} \ . \tag{6-8b}$$

For steel,

$$|Z_s| = 3.68 \times 10^{-5} \sqrt{f} \ . \tag{6-8c}$$

For any conductor, in general,

$$\boxed{|Z_s| = 3.68 \times 10^{-7} \sqrt{\frac{\mu_r}{\sigma_r}} \ \sqrt{f} \ .} \tag{6-8d}$$

Representative values of the relative permeability (μ_r) and the relative conductivity (σ_r) are listed in Table 6-1.

Table 6-1 Relative Conductivity and Permeability of Various Materials

Material	Relative conductivity σ_r	Relative permeability μ_r
Silver	1.05	1
Copper—annealed	1.00	1
Gold	0.70	1
Aluminum	0.61	1
Brass	0.26	1
Nickel	0.20	1
Bronze	0.18	1
Tin	0.15	1
Steel (SAE 1045)	0.10	1000
Lead	0.08	1
Monel	0.04	1
Stainless steel (430)	0.02	500

ABSORPTION LOSS

When an electromagnetic wave passes through a medium its amplitude decreases exponentially (Hayt, 1974) as shown in Fig. 6-6. This decay occurs because currents induced in the medium produce ohmic losses and heating of the material. Therefore, we can write

$$E_1 = E_0 e^{-t/\delta}, \tag{6-9}$$

and

$$H_1 = H_0 e^{-t/\delta}, \tag{6-10}$$

where $E_1(H_1)$ is the wave intensity at a distance t within the media, as shown in Fig. 6-6. The distance required for the wave to be attenuated to $1/e$ or 37% of its original value is defined as the skin depth, which is equal to

$$\delta = \sqrt{\frac{2}{\omega\mu\sigma}} \quad \text{meters.*} \tag{6-11a}$$

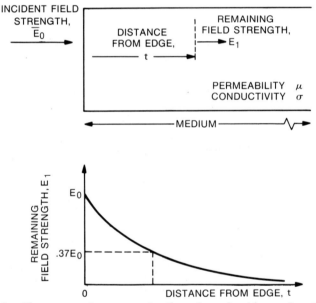

Figure 6-6. *Electromagnetic wave passing through an absorbing medium is attenuated exponentially.*

---

*Skin depth calculated by Eq. 6-11a is in meters when the constants listed on p. 142 (MKS system) are used.

To get an idea of typical skin depths for real materials, Eq. 6-11a can be revised. Substituting numerical values for μ and σ, and changing units so the depth is in inches gives

$$\delta = \frac{2.6}{\sqrt{f\mu_r\sigma_r}} \quad \text{in.} \tag{6-11b}$$

Some representative skin depths for copper, aluminum, and steel are listed in Table 6-2.

Table 6-2   Skin Depth of Various Materials

Frequency	δ For Copper (in.)	δ For Aluminum (in.)	δ For Steel (in.)
60 Hz	0.335	0.429	0.034
100 Hz	0.260	0.333	0.026
1 kHz	0.082	0.105	0.008
10 kHz	0.026	0.033	0.003
100 kHz	0.008	0.011	0.0008
1 MHz	0.003	0.003	0.0003
10 MHz	0.0008	0.001	0.00008

The absorption loss through a shield can now be written as

$$A = 20\left(\frac{t}{\delta}\right)\log(e) \quad \text{dB}, \tag{6-12a}$$

$$A = 8.69\left(\frac{t}{\delta}\right) \quad \text{dB}. \tag{6-12b}$$

As can be seen from the above equation, *the absorption loss in a shield one skin-depth thick is approximately 9 dB*. Doubling the thickness of the shield doubles the loss in dB.

Figure 6-7 is a plot of absorption loss in dB versus the ratio t/δ. This curve is applicable to plane waves, electric fields, or magnetic fields.

Substituting Eq. 6-11b into Eq. 6-12b gives the following expression for the absorption loss:

$$\boxed{A = 3.34t\sqrt{f\mu_r\sigma_r} \quad \text{dB.}} \tag{6-13}$$

In this equation, t is equal to the thickness of the shield in inches. Table 6-1 lists the relative conductivity and permeability of various materials that are likely to be used as shields.

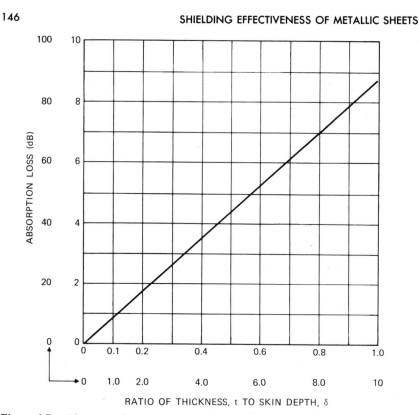

Figure 6-7. *Absorption loss is proportional to the thickness and inversely proportional to the skin depth of the medium. This plot can be used for electric fields, magnetic fields, or plane waves.*

Absorption loss versus frequency is plotted in Fig. 6-8 for two thicknesses of copper and steel. As can be seen, a thin (0.02 in.) sheet of copper provides significant absorption loss (66 dB) at 1 MHz but virtually no loss at frequencies below 1000 Hz. Figure 6-8 clearly shows the advantage of steel over copper in providing absorption loss. Even when steel is used, however, a thick sheet must be used to provide appreciable absorption loss below 1000 Hz.

REFLECTION LOSS

The reflection loss at the interface between two media is related to the difference in characteristic impedances between the media as shown in Fig. 6-9. The intensity of the transmitted wave from a medium with impedance

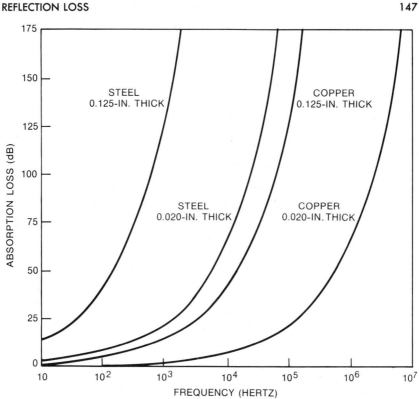

Figure 6-8. *Absorption loss increases with frequency and with shield thickness; steel offers more absorption loss than copper of the same thickness.*

Z_1 to a medium with impedance Z_2 (Hayt, 1974) is

$$E_1 = \frac{2Z_2}{Z_1 + Z_2} E_0,$$ (6-14)

and

$$H_1 = \frac{2Z_1}{Z_1 + Z_2} H_0.$$ (6-15)

$E_0(H_0)$ is the intensity of the incident wave, and $E_1(H_1)$ is the intensity of the transmitted wave.

When a wave passes through a shield, it encounters two boundaries, as shown in Fig. 6-10. The second boundary is between a medium with impedance Z_2 and a medium with impedance Z_1. The transmitted wave

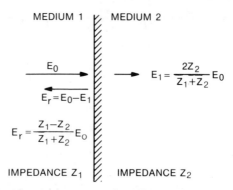

Figure 6-9. *An incident wave is partially reflected from, and partially transmitted through, an interface between two media. The transmitted wave is E_t and the reflected wave is E_r.*

$E_t(H_t)$ through this boundary is given by

$$E_t = \frac{2Z_1}{Z_1 + Z_2} E_1,$$ (6-16)

and

$$H_t = \frac{2Z_2}{Z_1 + Z_2} H_1.$$ (6-17)

If the shield is thick* compared to the skin depth, the total transmitted wave intensity is found by substituting Eqs. 6-14 and 6-15 into Eqs. 6-16 and 6-17, respectively. This neglects the absorption loss, which has been accounted for previously. Therefore, for thick shields the total transmitted wave is

$$E_t = \frac{4Z_1 Z_2}{(Z_1 + Z_2)^2} E_0,$$ (6-18)

and

$$H_t = \frac{4Z_1 Z_2}{(Z_1 + Z_2)^2} H_0.$$ (6-19)

Note that even though the electric and magnetic fields are reflected differently at each boundary, the net effect across both boundaries is the

---

*If the shield is not thick, multiple reflections occur between the two boundaries since the absorption loss in the shield is small. (See p. 155.)

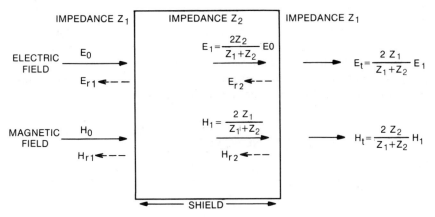

Figure 6-10. *Partial reflection and transmission occur at both faces of shield.*

same for both fields. If the shield is metallic and the surrounding area an insulator, then $Z_1 \gg Z_2$. Under these conditions, the largest reflection (smallest transmitted wave) occurs when the wave enters the shield (first boundary) for the case of electric fields, and when the wave leaves the shield (second boundary) for the case of magnetic fields. *Since the primary reflection occurs at the first surface in the case of electric fields, even very thin materials provide good reflection loss.* In the case of magnetic fields, however, the primary reflection occurs at the second surface and, as will be shown later, multiple reflections within the shield reduce the shielding effectiveness. When $Z_1 \gg Z_2$, Eqs. 6-18 and 6-19 reduce to

$$E_t = \frac{4Z_2}{Z_1} E_0, \qquad (6\text{-}20)$$

and

$$H_t = \frac{4Z_2}{Z_1} H_0. \qquad (6\text{-}21)$$

Substituting the wave impedance Z_w for Z_1, and the shield impedance Z_s for Z_2 the reflection loss for either the E or H field can be written as

$$R = 20 \log \frac{|Z_w|}{4|Z_s|} \quad \text{dB}, \qquad (6\text{-}22)$$

where

$Z_w =$ impedance of wave prior to entering the shield (Eq. 6-4),

$Z_s =$ impedance of the shield (Eq. 6-8d).

These reflection loss equations are for a plane wave approaching the interface at normal incidence. If the wave approaches at other than normal incidence, the reflection loss increases with the angle of incidence. The results also apply to other than plane waves, since any arbitrary field can be constructed from the superposition of plane waves. The results also apply to a curved interface, provided the radius of curvature is much greater than the skin depth.

REFLECTION LOSS TO PLANE WAVES

In the case of a plane wave (far field), the wave impedance Z_w equals the characteristic impedance of free space Z_0 (377 Ω). Therefore, Eq. 6-22 becomes

$$R = 20 \log \frac{94.25}{|Z_s|} \quad \text{dB.} \qquad (6\text{-}23\text{a})$$

Substituting Eq. 6-8d for $|Z_s|$ and rearranging Eq. 6-23a gives

$$\boxed{R = 168 - 10 \log(\,\mu_r f / \sigma_r) \quad \text{dB.}} \qquad (6\text{-}23\text{b})$$

The lower the shield impedance, the greater the reflection loss. Shield impedance is minimized by using a material with high conductivity and low permeability. Figure 6-11 is a plot of the reflection loss for three materials: copper, aluminum, and steel. Comparing this with Fig. 6-8 shows that although steel has more absorption loss than copper, it has less reflection loss.

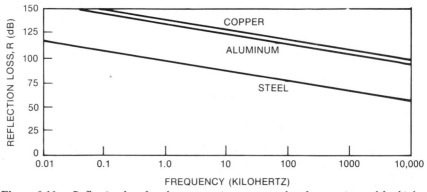

Figure 6-11. *Reflection loss for plane waves is greatest at low frequencies and for high conductivity material.*

COMPOSITE ABSORPTION AND REFLECTION LOSS TO PLANE WAVES

The total loss for plane waves in the far field is a combination of the absorption and reflection losses, as indicated in Eq. 6-3. The multiple reflection correction term B is normally neglected for plane waves, since the reflection loss is so high and the correction term is small. If the absorption loss is greater than 1 dB, the correction term is less than 11 dB; if the absorption loss is greater than 4 dB, the correction is less than 2 dB.

Figure 6-12 shows the overall attenuation or shielding effectiveness of a 0.020-in. thick copper shield. As can be seen, the reflection loss decreases with increasing frequency; this is because the shield impedance Z_s increases with frequency. The absorption loss, however, increases with frequency, due to the decreasing skin depth. The minimum shielding effectiveness occurs at some intermediate frequency, in this case at 10 kHz. From Fig. 6-12, it is apparent that for low frequency plane waves, reflection loss accounts for most of the attenuation, whereas most of the attenuation at high frequencies comes from absorption loss.

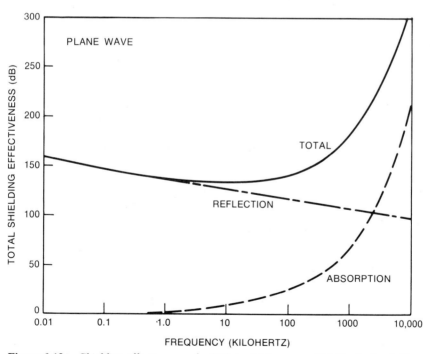

Figure 6-12.   *Shielding effectiveness of a 0.02-in. thick copper shield in the far field.*

REFLECTION LOSS IN THE NEAR FIELD

In the near field, the ratio of the electric field to the magnetic field is no longer determined by the characteristic impedance of the medium. Instead, the ratio of the electric field to the magnetic field depends more on the characteristics of the source (antenna). If the source has high voltage and low current, the wave impedance is greater than 377 Ω and the field will be a high-impedance, or electric field. If the source has low voltage and high current, the wave impedance will be less than 377 Ω and the field will be a low-impedance, or magnetic field.

Since the reflection loss (Eq. 6-22) is a function of the ratio between the wave impedance and the shield impedance, the reflection loss varies with the wave impedance. A high-impedance (electric) field, therefore, has higher reflection loss than a plane wave. Similarly, a low-impedance (magnetic) field has lower reflection loss than a plane wave. This is shown in Fig. 6-13 for a copper shield separated from the source by distances of 1 and 30 m. Also shown for comparison is the plane wave reflection loss.

For any specified distance between source and shield, the three curves

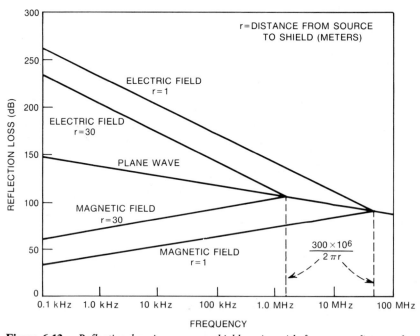

Figure 6-13. *Reflection loss in a copper shield varies with frequency, distance from the source, and type of wave.*

(electric field, magnetic field, and plane wave) of Fig. 6-13 merge at the frequency that makes the separation between source and shield equal to $\lambda/2\pi$. When the spacing is 30 m, the electric and magnetic field curves come together at a frequency of 1.6 MHz.

The curves shown in Fig. 6-13 are for point sources producing only an electric field or only a magnetic field. Most practical sources, however, are a combination of both electric and magnetic fields. The reflection loss for a practical source, therefore, lies somewhere between the electric field lines and the magnetic field lines shown in the figure.

Figure 6-13 shows that the reflection loss of an electric field decreases with frequency until the separation distance becomes $\lambda/2\pi$. Beyond that, the reflection loss is the same as for a plane wave. The reflection loss of a magnetic field increases with frequency, again until the separation distance becomes $\lambda/2\pi$. Then the loss begins to decrease at the same rate as that of a plane wave.

ELECTRIC FIELD REFLECTION LOSS

The wave impedance due to a point source of electric field can be approximated by the following equation when $r < \lambda/2\pi$

$$|Z_w|_e = \frac{1}{2\pi f \epsilon r}, \qquad (6\text{-}24)$$

where r is the distance from the source to the shield in meters and ϵ is the dielectric constant. The reflection loss can be determined by substituting Eq. 6-24 into Eq. 6-22, giving

$$R_e = 20 \log \frac{1}{8\pi f \epsilon r |Z_s|} \quad \text{dB}, \qquad (6\text{-}25)$$

or

$$R_e = 20 \log \frac{4.5 \times 10^9}{f r |Z_s|} \quad \text{dB}, \qquad (6\text{-}26)$$

where r is in meters. The characteristic impedance of the shield Z_s can be determined from Eq. 6-8d.

In Fig. 6-13 the lines labeled "electric field" are plots of Eq. 6-26 for a copper shield with r equal to 1 and 30 m. The equation and the plot represent the reflection loss at a specified distance from a point source producing only an electric field. An actual electric field source, however, has some small magnetic field component in addition to the electric field. It therefore has a reflection loss somewhere between the electric field line and the plane wave line of Fig. 6-13. Since, in general, we do not know where between these two lines the actual source may fall, the plane wave calculations (Eq. 6-23) are normally used in determining the reflection loss for an electric field. The actual reflection loss is then equal to or greater than that calculated in Eq. 6-23.

COMPOSITE ABSORPTION AND REFLECTION LOSS FOR ELECTRIC FIELDS

The total loss for an electric field is obtained by combining the absorption (Eq. 6-13) and reflection losses (Eq. 6-23), as indicated in Eq. 6-3. The multiple reflection correction factor B is normally neglected in the case of an electric field since the reflection loss is so great and the correction term is small. *Reflection loss is the primary shielding mechanism for electric fields*.

MAGNETIC FIELD REFLECTION LOSS

The wave impedance due to a point source of magnetic field can be approximated by the following equation assuming $r < \lambda/2\pi$

$$|Z_w|_m = 2\pi f \mu r, \tag{6-27}$$

where r is the distance from the source to the shield and μ is the permeability. The reflection loss can be determined by substituting Eq. 6-27 into Eq. 6-22, giving

$$R_m = 20 \log \frac{2\pi f \mu r}{4|Z_s|} \quad \text{dB}, \tag{6-28}$$

or

$$R_m = 20 \log \frac{1.97 \times 10^{-6} fr}{|Z_s|} \quad \text{dB}, \tag{6-29a}$$

where r is in meters. Substituting Eq. 6-8d for $|Z_s|$ and rearranging Eq. 6-29a gives

$$R_m = 14.6 + 10 \log\left(\frac{fr^2 \sigma_r}{\mu_r}\right) \quad \text{dB,}^* \tag{6-29b}$$

with r in meters.

*If a negative value is obtained in the solution for R, use $R = 0$ instead and neglect the multiple reflection factor B. If a solution for R is positive and near zero, Eq. 6-29 is slightly in error. The error occurs because the assumption $Z_1 \gg Z_2$, made during the derivation of the equation, is not satisfied in this case. The error is 3.8 dB when R equals zero, and it decreases as R gets larger. From a practical point of view, however, even this error can be neglected.

In Fig. 6-13 the curves labeled "magnetic field" are plots of Eq. 6-29 for a copper shield with r equal to 1 and 30 m. Equation 6-29 and the plot in Fig. 6-13 represent the reflection loss at the specified distance from a point source producing only a magnetic field. Most real magnetic field sources have a small electric field component in addition to the magnetic field and the reflection loss lies somewhere between the magnetic field line and the plane wave line of Fig. 6-13. Since we do not generally know where between these two lines the actual source may fall, Eq. 6-29 should be used to determine the reflection loss for a magnetic field. The actual reflection loss will then be equal to or greater than that calculated in Eq. 6-29.

Where the distance to the source is not known, the near field magnetic reflection loss can usually be assumed to be zero at low frequencies.

MULTIPLE REFLECTIONS IN THIN SHIELDS

If the shield is thin, the reflected wave from the second boundary is re-reflected off the first boundary and again returns to the second boundary to be reflected, as shown in Fig. 6-14. This can be neglected in the case of a thick shield, since the absorption loss is high. By the time the wave reaches the second boundary for the second time, it is of negligible amplitude, since by then it has passed through the thickness of the shield three times.

For electric fields, most of the incident wave is reflected at the first boundary and only a small percentage enters the shield. This can be seen from Eq. 6-14 and the fact that $Z_2 \ll Z_1$. Therefore, multiple reflections within the shield can be neglected for electric fields.

For magnetic fields most of the incident wave passes into the shield at

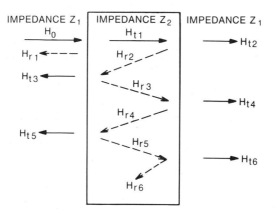

Figure 6-14. *Multiple reflections occur in a thin shield; part of the wave is transmitted through at each reflection.*

the first boundary, as shown in Eq. 6-15 when $Z_2 \ll Z_1$. The magnitude of the transmitted wave is actually double that of the incident wave. With such a large magnitude magnetic field within the shield, the effect of multiple reflections inside the shield must be considered.

The multiple reflection correction factor for magnetic fields in a shield of thickness t and skin depth δ is

$$B = 20 \log(1 - e^{-2t/\delta}) \quad \text{dB.} \qquad \text{*} \qquad (6\text{-}30)$$

Figure 6-15 is a plot of the correction factor B as a function of t/δ. Note that the correction factor is a negative number, indicating that less shielding is obtained from a thin shield due to reflection.

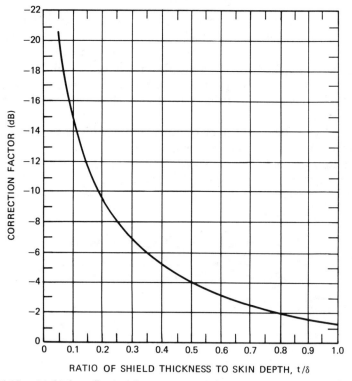

Figure 6-15. *Multiple reflection loss correction factor (B) for thin shields, with magnetic fields.*

*See Appendix C for this calculation.

COMPOSITE ABSORPTION AND REFLECTION LOSS FOR MAGNETIC FIELDS

The total loss for a magnetic field is obtained by combining the absorption loss (Eq. 6-13) and the reflection loss (Eq. 6-29), as indicated in Eq. 6-3. If the shield is thick (absorption loss > 10 dB) the multiple reflection correction factor B can be neglected. If the shield is thin the correction factor from Eq. 6-30 or Fig. 6-15 must be included.

In the near field, the reflection loss to a low frequency magnetic field is small. Due to multiple reflections this effect is even more pronounced in a thin shield. *The primary loss for low frequency magnetic fields is, therefore, absorption loss.* Additional protection against low frequency magnetic fields can be achieved only by providing a low reluctance magnetic shunt path to divert the field around the circuit being protected. This is shown in Fig. 6–16.

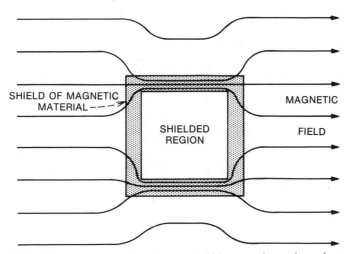

Figure 6-16. *Magnetic material used as a shield by providing a low-reluctance path for the magnetic field, diverting it around the shielded region.*

SUMMARY OF SHIELDING EQUATIONS

Figure 6–17 is a summary showing which equations are used to determine shielding effectiveness under various conditions. A qualitative summary of the shielding provided by solid shields under various conditions is given in the summary at the end of this chapter (p. 171).

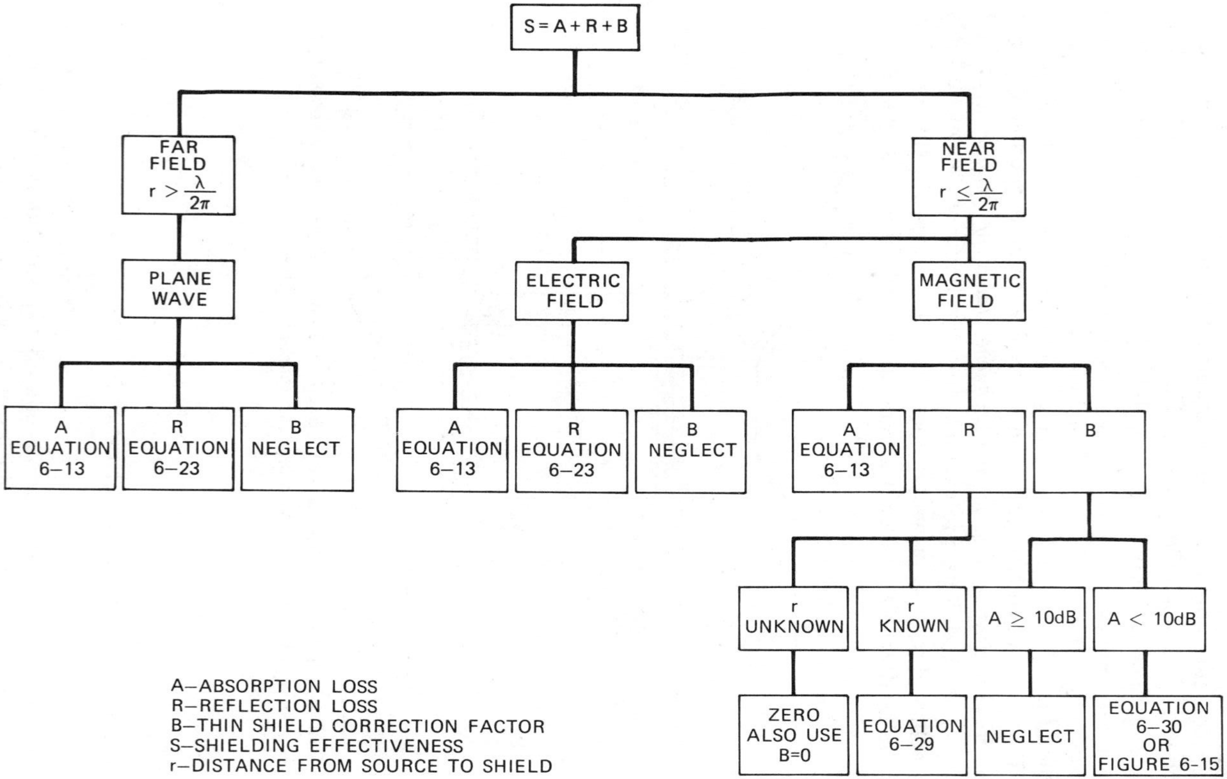

Figure 6-17. Shielding effectiveness summary shows which equations are used to calculate shielding effectiveness under various conditions.

MAGNETIC MATERIAL AS A SHIELD

If a magnetic material is used as a shield in place of a good conductor, there is an increase in the permeability μ and a decrease in the conductivity σ. This has the following effects:

1. It increases the absorption loss, since the permeability increases more than the conductivity decreases for most magnetic materials. (See Eq. 6–13.)

2. It decreases the reflection loss, since the shield impedance Z_s increases. (See Eq. 6–23b.) The total loss through a shield is the sum of that due to absorption and that due to reflection. *In the case of low frequency magnetic fields there is very little reflection loss, and absorption loss is the primary shielding mechanism. Under these conditions it is advantageous to use a magnetic material to increase the absorption loss.* In the case of low frequency electric fields or plane waves, the primary shielding mechanism is reflection. In this case, using a magnetic material would decrease the shielding.

When magnetic materials are considered as a shield, three often overlooked properties must be taken into account. These are:

1. Decrease of permeability with frequency.

2. Permeability depends on field strength.

3. Machining or working high permeability magnetic materials, such as mumetal, may change their magnetic properties.

Most permeability values given for magnetic materials are static, or dc, permeabilities. As frequency increases, the permeability decreases. Usually, the larger the dc permeability, the greater the decrease with frequency. Figure 6–18 plots permeability against frequency for a variety of magnetic materials. As can be seen, mumetal is no better than cold rolled steel at 100 kHz, even though the dc permeability is thirteen times that of cold rolled steel. High permeability materials are most useful as magnetic field shields at frequencies below 10 kHz.

The usefulness of magnetic materials as a shield varies with the field strength H. A typical magnetization curve is shown in Fig. 6–19. The static permeability is the ratio of B to H. As can be seen, maximum permeability, and therefore shielding, occur at a medium level of field strength. At both higher and lower field strengths the permeability, and hence the shielding, is lower. The effect at high field strengths is due to saturation, which varies depending on the type of material and its thickness. At field strengths well above saturation, the permeability falls off rapidly. In general, the higher the permeability the lower the field strength that causes saturation. Most magnetic material specifications give the best permeability, namely that at optimum frequency and field strength. Such specifications can be very misleading.

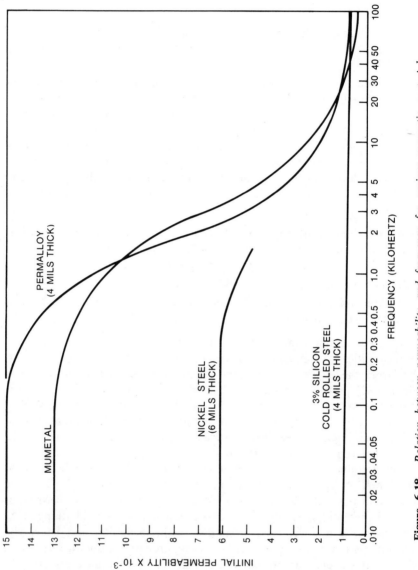

Figure 6-18. *Relation between permeability and frequency for various magnetic materials.*

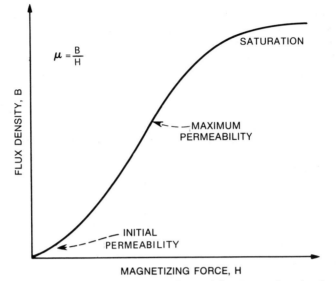

Figure 6-19. *Typical magnetization curve. Permeability is equal to the slope of the curve.*

To overcome the saturation phenomenon, multilayer magnetic shields can be used. An example is shown in Fig. 6–20. There, the first shield (a low permeability material) saturates at a high level, and the second shield (a high permeability material) saturates at a low level. The first shield reduces the magnitude of the magnetic field to a point that does not saturate the second; the second shield then provides the majority of the magnetic field shielding. These shields can also be constructed using a conductor, such as copper, for the first shield, and a magnetic material for

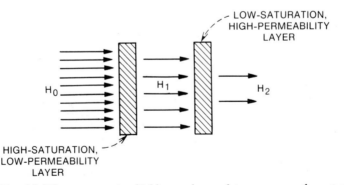

Figure 6-20. *Multilayer magnetic shields can be used to overcome the saturation phenomenon.*

the second. The low permeability, high saturation material is always placed on the side of the shield closest to the source of the magnetic field. In some difficult cases, additional shield layers may be required to obtain the desired field attenuation. Another advantage of multilayer shields is that there is increased reflection loss due to the additional reflecting surfaces.

Machining or working of some high permeability materials, such as mumetal or permalloy, may degrade their magnetic properties. This can also happen if the material is dropped or subjected to shock. The material must then be properly annealed, in order to restore its magnetic properties.

EXPERIMENTAL DATA

Results of tests performed to measure the magnetic field shielding effectiveness of various types of metallic sheets are shown in Fig. 6–21 and 6–22. The measurements were made in the near field with the source and

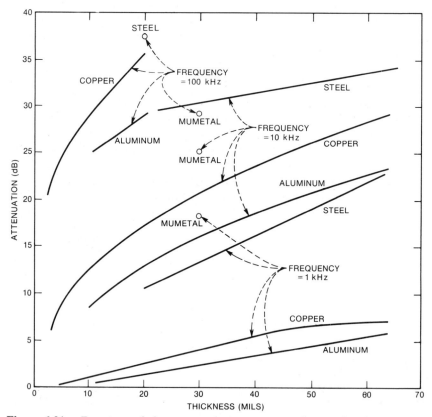

Figure 6-21. *Experimental data on magnetic attenuation by metallic sheets in the near field.*

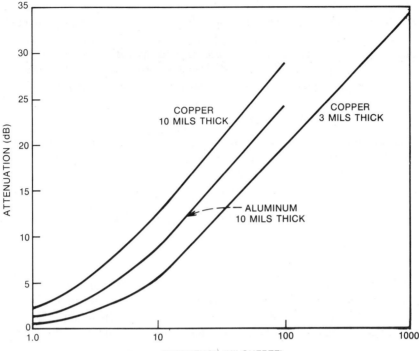

Figure 6-22. *Experimental results of tests to determine magnetic field attenuation of conducting sheets in the near field.*

receiver 0.1-in. apart. The shields were from 3 to 60 mils (in. $\times 10^{-3}$) thick, and the test frequency ranged from 1 to 100 kHz. Figure 6–21 clearly shows the superiority of steel over copper for shielding magnetic fields at 1 kHz. But at 100 kHz, steel is only slightly better than copper. Somewhere between 100 kHz and 1 MHz, however, a point is reached where copper becomes a better shield than steel.

Figure 6–21 also demonstrates the effect of frequency on mumetal as a magnetic shield. At 1 kHz, mumetal is more effective than steel, but at 10 kHz steel is more effective than mumetal. At 100 kHz, steel, copper, and aluminum are all better than mumetal.

In Fig. 6–22 some of the data from Fig. 6–21 is replotted to show the attenuation provided by thin copper and aluminum shields at various frequencies.

In summary, a magnetic material such as steel or mumetal makes a better magnetic field shield at low frequencies than does a good conductor such as aluminum or copper. At high frequencies, however, the good conductors provide the better magnetic shielding.

The magnetic shielding effectiveness of solid, nonmagnetic shields increases with frequency. Therefore, measurements of shielding effectiveness should be made at the lowest frequency of interest. The shielding effectiveness of magnetic materials may decrease with increasing frequency due to the decreasing permeability. The effectiveness of nonsolid shields may also decrease with frequency due to the increased leakage through the holes.

SEAMS AND HOLES

The previous calculations of shielding effectiveness have assumed a solid shield with no seams or holes. It has been shown that with the exception of low frequency magnetic fields, it is quite easy to obtain more than 90 dB of shielding effectiveness. In practice, however, most shields are not solid. There must be access covers, doors, holes for conductors, ventilation, switches, meters, and mechanical joints and seams. All of these may considerably reduce the effectiveness of the shield. *As a practical matter, the intrinsic shielding effectiveness of the material is of less concern than the leakage through seams, joints, and holes.*

Shield discontinuities usually have more effect on magnetic field leakage than on electric field leakage. Accordingly, greater emphasis is given to methods of minimizing the magnetic field leakage. In almost all cases, these same methods are more than adequate for minimizing the electric field leakage.

The amount of leakage from a shield discontinuity depends mainly on three items:

1. The maximum linear dimension (not area) of the opening,
2. The wave impedance,
3. The frequency of the source.

The fact that maximum dimension, not area, determines the amount of leakage can best be visualized by considering the circuit theory approach to shielding. In that approach, the noise fields induce currents in the shield, and these currents then generate additional fields. The new fields cancel the original field in some regions of space. For this cancellation to occur, these shield currents must be allowed to flow undisturbed in the manner in which they were induced by the incident field. If a shield discontinuity forces the induced currents to flow in a different path, the shielding effectiveness is reduced. The further the current is forced to detour, the greater will be the decrease in shielding effectiveness.

Figure 6–23 shows how discontinuities affect the induced shield currents. Figure 6–23A shows a section of shield containing no discontinuity. Also shown are the induced shield currents. Figure 6–23B shows how a rectangular slot detours the induced shield currents, and hence produces leakage. Figure 6–23C shows a much narrower slit of the same length. This narrower slit has almost the same effect on the current as the wider slot of Fig. 6–23B and therefore produces the same amount of leakage. Figure 6–23D shows that a group of small holes has much less detouring effect on the current than the slot of Fig. 6–23B, and therefore, produces less leakage even if the total area is the same as the slot. From this it should be obvious that *a large number of small holes produce less leakage than a large hole of the same total area.*

A rectangular slot as shown in Fig. 6–23B or C forms a slot antenna. Such an antenna, even if very narrow, can cause considerable leakage if it is longer than 1/100 wavelength. Seams and joints often form very efficient slot antennas. Maximum radiation occurs from a slot antenna when the length is equal to one half wavelength.

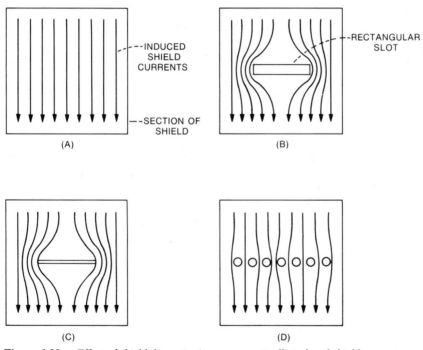

Figure 6-23. *Effect of shield discontinuity on magnetically induced shield current.*

Waveguide Below Cutoff

Additional attenuation can be obtained from a hole if it is shaped to form a waveguide, as shown in Fig. 6–24. A waveguide has a cutoff frequency below which it becomes an attenuator. The attenuation is a function of the length t of the waveguide. For a round waveguide, the cutoff frequency is

$$f_c = \frac{6.9 \times 10^9}{d} \quad \text{Hz,} \tag{6-31}$$

where d is the diameter in inches. For a rectangular waveguide the cutoff frequency is

$$f_c = \frac{5.9 \times 10^9}{l} \quad \text{Hz,} \tag{6-32}$$

where l is the largest dimension of the waveguide cross section in inches.

As long as the operating frequency is much less than the cutoff frequency, the magnetic field shielding effectiveness of a round waveguide (Quine, 1957) is

$$S = 32 \frac{t}{d} \quad \text{dB,} \tag{6-33}$$

where d is the diameter and t is the length of the waveguide, as shown in Fig. 6–24. For a rectangular waveguide (Quine, 1957)

$$S = 27.2 \frac{t}{l} \quad \text{dB,} \tag{6-34}$$

where l is the largest linear dimension of the waveguide cross section and t is the length. A waveguide having a length three times its diameter provides greater than 100 dB of shielding.

If a hole in a shield has a diameter less than the shield thickness, a waveguide is formed. In this case the length of the waveguide is equal to the thickness of the shield.

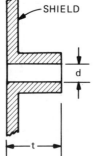

Figure 6-24. *Cross section of a hole formed into a waveguide with diameter d and length t.*

Round Holes

A common way to provide ventilation is to use the configuration shown in Fig. 6–25. This figure shows a section of shield containing a square array of round holes. The hole diameter is d, the hole center-to-center spacing is c, and the overall dimension of the array is l. Quine (1957) has shown that the magnetic field shielding effectiveness* is

$$S = 20 \log \frac{c^2 l}{d^3} + 32 \frac{t}{d} + 3.8 \quad \text{dB.} \tag{6-35}$$

This equation shows the shielding effectiveness independent of frequency and applies provided d is less than $\lambda/2\pi$. For a rectangular array of outside dimensions l_1 and l_2, use $l = \sqrt{l_1 l_2}$ in Eq. 6–35. The first term in this equation represents the leakage through holes in a thin shield. This term shows the shielding effectiveness to be inversely proportional to the cube of the hole diameter, and directly proportional to the square of the hole center-to-center spacing. The second term is a thickness correction factor obtained by treating each hole as a waveguide below cutoff.

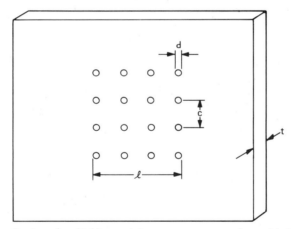

Figure 6-25. *Section of a shield containing a square array of round holes.*

CONDUCTIVE GASKETS

Joints made with continuous welding or brazing provide the maximum shielding. Rivets and screws make less desirable joints. If screws are used, they should be as close together as practical. Every attempt should be

---

*Shielding effectiveness in this case is the increased attenuation provided by the hole pattern over that provided if the total area ($l \times l$) had been removed from the shield.

made to maintain electrical continuity across the joint to avoid forming a slot antenna. It may be desirable to use EMI gaskets on joints. These are conductive gaskets which, when properly compressed, provide electrical continuity across a joint. They are capable of controlling leakage at frequencies from the low kilohertz up to tens of gigahertz.

One of the most common types of EMI gaskets is made of a knitted wire mesh. They are available in strips, with rectangular or round cross sections, or in preformed shapes. The gaskets are available in various materials, including steel-copper alloy, silver-plated brass, aluminum, and monel. The gasket material used should be galvanically compatible with the mating surface to minimize corrosion. For this reason, monel and silver-plated brass should not normally be used with an aluminum enclosure.

Figure 6–26 shows the correct and incorrect way to install an EMI gasket between an enclosure and its cover. The gasket should be in a slot and on the inside of the screw to protect against leakage around the screw hole. For electrical continuity across the joint or seam, the metal should be free of paints, oxides, and insulating films. The metal should be protected from corrosion with a conductive finish. Do not anodize aluminum; rather, use an alodine or chromate finish, both of which are conductive.

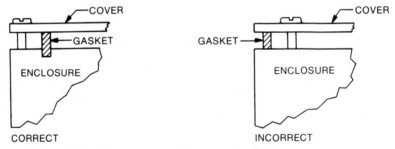

Figure 6-26. *EMI gaskets, correct and incorrect installation.*

If both EMI protection and environmental protection are required, two separate gaskets or a combination EMI and environmental gasket may be used. The combination gasket usually has a knitted wire mesh mated to silicone rubber. If both environmental and EMI gaskets are installed either as a combination unit or as two separate gaskets, the EMI gasket should be on the inside of the environmental gasket.

With a sheet metal enclosure, the EMI gasket may be mounted by one of the methods shown in Fig. 6–27.

Perforated sheet stock or screening should be used to cover ventilation openings. The material must have electrical continuity between the strands where they cross. The entire perimeter of the screening must be in electrical contact with the chassis.

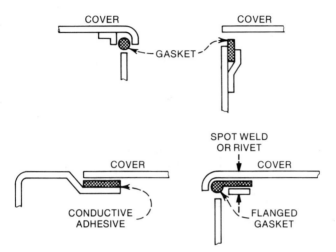

Figure 6-27.   *Suitable ways to install EMI gaskets in sheet metal enclosures.*

Conductive gaskets can be used around switches and controls mounted in the shield. These should be mounted as shown in Fig. 6–28. Large holes cut in panels for meters can completely destroy the effectiveness of a shield. If meters are used in a shield panel, they should be mounted as

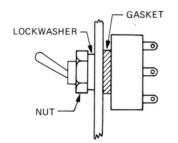

Figure 6-28.   *Switch mounted in a panel using an EMI gasket.*

shown in Fig. 6–29 in order to provide shielding of the meter hole. Wires entering the shield should be filtered as explained in Chapter 4. Shield enclosures should be electrically grounded.

For optimum shielding the enclosure should be thought of as "electrically watertight" with EMI gaskets used in place of normal environmental gaskets.

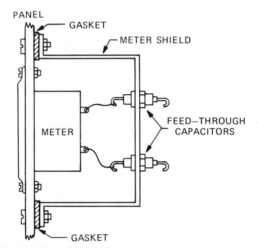

Figure 6-29.   *Method of shielding a meter hole in a panel.*

SUMMARY

- Reflection loss is very large for electric fields and plane waves.
- Reflection loss is normally small for low frequency magnetic fields.
- A shield one skin depth thick provides approximately 9 dB of absorption loss.
- Magnetic fields are harder to shield against than electric fields.
- Use a magnetic material to shield against low frequency magnetic fields.
- Use a good conductor to shield against electric fields, plane waves, and high-frequency magnetic fields.
- Actual shielding effectiveness obtained in practice is usually determined by the leakage at seams and joints, not by the shielding effectiveness of the material itself.
- The maximum dimension (not area) of a hole or discontinuity determines the amount of leakage.
- A large number of small holes result in less leakage than a larger hole of the same total area.

- The following is a qualitative summary of shielding (solid shield, no holes or seams):

Material	Frequency (kHz)	Absorption loss[a] all fields	Reflection loss		
			Magnetic field[b]	Electric field	Plane wave
Magnetic	< 1	Bad–Poor	Bad	Excellent	Excellent
($\mu_r = 1000$,	1–10	Average–Good	Bad–Poor	Excellent	Excellent
$\sigma_r = 0.1$)	10–100	Excellent	Poor	Excellent	Good
	> 100	Excellent	Poor–Average	Good	Average–Good
Non	< 1	Bad	Poor	Excellent	Excellent
magnetic	1–10	Bad	Average	Excellent	Excellent
($\mu_r = 1$,	10–100	Poor	Average	Excellent	Excellent
$\sigma_r = 1$)	> 100	Average–Good	Good	Excellent	Excellent

Key	Attenuation
Bad	0–10 dB
Poor	10–30 dB
Average	30–60 dB
Good	60–90 dB
Excellent	> 90 dB

[a]Absorption Loss for 1/32-in. thick shield.

[b]Magnetic field reflection loss for a source distance of 1 m. (Shielding is less if distance is less than 1 m and more if distance is greater than 1 m.)

BIBLIOGRAPHY

Bardell, P. R., *Magnetic Materials in the Electrical Industry*, Macdonald and Co., London, 1960.

Burgoon, J. R., Jr., "Fundamentals of Electrical Shield Design," *Insulation / Circuits*, August, 1970.

Cook, D. V., "RFI Suppression, Part I," *Electromechanical Design*, Vol. 11, November, 1967.

Cowdell, R. B., "Nomographs Simplify Calculations of Magnetic Shielding Effectiveness," *EDN*, Vol. 17, September 1, 1972.

Ficchi, R. O., *Electrical Interference*, Hayden Book Co., New York, 1964.

Ficchi, R. O., *Practical Design for Electromagnetic Compatibility*, Hayden Book Co., New York, 1971.

Frederick Research Corp, *Handbook on Radio Frequency Interference*, Vol. 3 (Methods of Electromagnetic Interference Suppression), Frederick Research Corp., Wheaton, Maryland, 1962.

Hayt, W. H., Jr., *Engineering Electromagnetics*, Third Edition, McGraw-Hill, New York, 1974.

Jordan, E. C., and Balmain, K. G., *Electromagnetic Waves and Radiating Systems*, Second Edition, Prentice-Hall, Englewood Cliffs, N. J., 1968.

McHenry, G. A., "Audio Frequency Interference Considerations in Aircraft Electronic Systems Design," *Proceedings of the Conference on Radio Interference Reduction*, Armour Research Foundation, pp. 121–133, December, 1954.

Miller, D. A., and Bridges, J. E., "Review of Circuit Approach to Calculate Shielding Effectiveness," *IEEE Transactions on Electromagnetic Compatibility*, Vol. EMC-10, March, 1968.

Quine, J. P., "Theoretical Formulas for Calculating the Shielding Effectiveness of Perforated Sheets and Wire Mesh Screens," *Proceedings of the Third Conference on Radio Interference Reduction*, Armour Research Foundation, pp. 315–329, February, 1957.

Severinsen, J., "Designers Guide to EMI Shielding," Parts 1 and 2, *EDN*, Vol. 20, February 5, 1975 and March 20, 1975.

Vasaka, C. S., "Problems in Shielding Electronic Equipment," *Proceedings of the Conference on Radio Interference Reduction*, Armour Research Foundation, pp. 86–103, December, 1954.

Weeks, W. L., *Electromagnetic Theory for Engineering Applications*, Wiley, New York, 1964.

White, D. R. J., *Electromagnetic Interference and Compatibility*, Vol. 3 (EMI Control Methods and Techniques), Don White Consultants, Germantown, Maryland, 1973.

Young, F. J., "Ferromagnetic Shielding Related to the Physical Properties of Iron," *1968 IEEE Electromagnetic Compatibility Symposium Record*, IEEE, New York, 1968.

7 CONTACT PROTECTION

Whenever contacts open or close a current-carrying circuit, electrical breakdown may develop between the contacts. This breakdown begins while the contacts are close together, but not quite touching. In the case of contacts that are closing, the breakdown continues until the contacts are closed. In the case of contacts that are opening, the breakdown continues until conditions can no longer support the breakdown. Whenever breakdown occurs, some physical damage is done to the contacts, decreasing their useful life. In addition, breakdown can also lead to high frequency radiation, and to voltage and current surges in the wiring. These surges may be the source of interference affecting other circuits.

The techniques used to minimize physical damage to the contacts are similar to those used to eliminate radiated and conducted interference. All of the contact protection networks discussed in this chapter greatly reduce the amount of noise generated by the contacts and the load, as well as extend the life of the contacts. Two types of breakdown are important in switching contacts. They are the gas or glow discharge and the metal-vapor or arc discharge.

GLOW DISCHARGES

A regenerative, self-supporting, glow discharge can occur between two contacts when the gas between the contacts becomes ionized. Such a breakdown is also called a Townsend discharge. The voltage necessary to initiate a glow discharge is a function of the gas, the spacing of the contacts, and the gas pressure. If the gas is air at standard temperature and pressure, 320 V is required at a gap length of 0.0003 in. to initiate a glow discharge. If the gap is shorter or longer, more voltage is required. Figure 7-1 shows the required breakdown voltage (V_B) for starting the glow discharge versus the separation distance of the contacts. After the breakdown occurs, a somewhat smaller voltage (V_G) is sufficient to keep the gas ionized. In air V_G is approximately 300 V. As can be seen in Fig. 7-1 this sustaining voltage is nearly constant regardless of the contact spacing. A minimum current is also necessary to sustain the glow; typically, this is a few milliamperes.

To avoid a glow discharge, the voltage across the contacts should be kept below 300 V. If this is done, the only concern is contact damage due to an arc discharge.

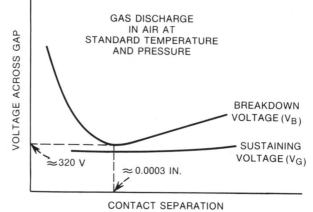

Figure 7-1. *Voltage versus distance for glow discharge.*

METAL-VAPOR OR ARC DISCHARGES

An arc discharge can occur at contact spacings and voltages much below those required for a glow discharge. It can even occur in a vacuum, since it does not require the presence of a gas. An arc discharge is started by field-induced electron emission, which requires a voltage gradient of approximately 0.5 MV/cm (5 V at 4×10^{-6} in.).

An arc is formed whenever an energized but unprotected contact is opened or closed, since the voltage gradient usually exceeds the required value when the contact spacing is small. When the arc discharge forms, the electrons emanate from a small area of the cathode—where the electric field is strongest.

Since, on a microscopic scale, all surfaces are rough, the highest and sharpest point on the cathode has the largest voltage gradient and becomes the source of electrons for the field emission. This is shown in Fig. 7-2. The electron stream fans out as it crosses the gap and finally bombards the anode. The localized current has a very high density, and it heats the contact material (due to I^2R losses) to a few thousand degrees Kelvin. This may be enough to vaporize the contact metal. In general, either the anode or cathode may vaporize first, depending on the rates at which heat is delivered to and removed from the two contacts. This in turn depends upon the size, material, and spacing of the two contacts.

The appearance of molten metal marks the transition from field emission (electron flow) to a metal-vapor arc. This transition typically takes place in times less than a nanosecond. The molten metal, once present,

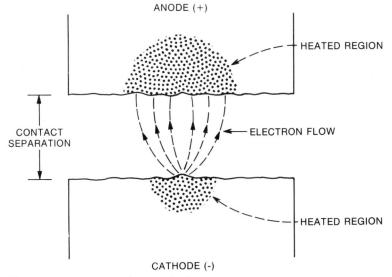

Figure 7-2. *Initiation of an arc discharge.*

forms a conductive "bridge" between the contacts, thus maintaining the arc even though the voltage gradient may have decreased below the value necessary to initiate the discharge. This metal vapor bridge draws a current limited by the supply voltage and the impedance of the circuit. After the arc has started, it persists as long as the external circuit provides enough voltage to overcome the cathode contact potential and enough current to vaporize the anode or the cathode material. As the contacts continue to separate, the molten metal "bridge" stretches and eventually ruptures. The minimum voltage and current required to sustain the arc are called the minimum arcing voltage (V_A) and the minimum arcing current (I_A). Typical values of minimum arcing voltage and current are shown in Table 7-1 (National Association of Relay Manufacturers, 1969). If either the voltage or current falls below these values the arc is extinguished.

For arcs between contacts of different materials, V_A is determined by the cathode (negative contact) material, and I_A is assumed to be whichever contact material (anode or cathode) has the lowest arcing current. Note, however, that the minimum arcing currents listed in Table 7-1 are for clean, undamaged contacts. After the contacts have become damaged from some arcing, the minimum arcing current may decrease to as low as one tenth of the value listed in the table.

Table 7-1   Contact Arcing Characteristics

Material	Minimum arcing voltage (V_A)	Minimum arcing current (I_A)
Silver	12	400 mA
Gold	15	400 mA
Gold alloy[a]	9	400 mA
Palladium	16	800 mA
Platinum	17.5	700 mA

[a]69% gold, 25% silver, 6% platinum.

In summary, an arc discharge is a function of the contact material, and it is characterized by a relatively low voltage and a high current. In contrast, a glow discharge is a function of the gas, usually air, between the contacts, and it is characterized by a relatively high voltage and low current. As will be seen in a later section, it is difficult to prevent an arc discharge from forming, since only a low voltage is required. If the arc does form, however, it should be prevented from sustaining itself by keeping the available current below the minimum arcing current.

AC VERSUS DC CIRCUITS

If the contact is to survive, the arc, once formed, must be broken rapidly to minimize damage to the contact material. If it is not broken rapidly enough, some metal transfers from one contact to the other. The damage done by an arc is proportional to the energy in it—namely (Voltage)× (Current)×(Time).

The higher the voltage across the contacts, the more difficult it is to interrupt the arc. Under arcing conditions, a set of contacts can usually handle their rated number of volt-amperes at a voltage equal to or less than the rated voltage, but not necessarily at a higher voltage.

A set of contacts can normally handle a much higher ac than dc voltage, for the following reasons:

1. The average value of an ac voltage is less than the rms value.

2. During the time that the voltage is less than 10–15 V, an arc is very unlikely to start.

3. Due to polarity reversal, each contact is an anode and a cathode an equal number of times.

4. The arc will be extinguished when the voltage goes through zero.

A contact rated at 30 V dc can, therefore, typically handle 115 V ac. One disadvantage of switching ac, however, is that it is much harder to provide adequate contact protection networks when they are required.

CONTACT MATERIAL

Various load levels (currents) require different types of contact materials. No one material is useful from zero current (dry circuit) up to high current. Palladium is good for high-current loads under eroding contact conditions. Silver and silver cadmium operate well at high current but may fail under conditions of no arcing. Gold and gold alloys work well under low-level or dry-circuit conditions but erode excessively at high currents.

Many so-called "general purpose relays" are on the market, rated from dry circuit to 2 A. These are usually made by plating hard gold over a heavy load contact material such as silver or palladium. When used for low current, the contact resistance remains low due to the gold plating. When used for high load current, the gold is burned off during the first few operations, and the high current contact material remains. For this reason, once a general purpose relay is used with high currents it is no longer usable in a low current application.

A problem sometimes occurs when soft gold is plated over silver. The silver migrates through the gold and forms a high resistance coating (silver-sulfide) on the contact. This may then cause the contact to fail due to the high resistance surface coating.

CONTACT RATING

Contacts are normally rated by the maximum values of voltage and current they can handle feeding a resistive load. When a contact is operated at its rated conditions, there is some momentary arcing on "make" and "break".* When operated under these conditions a contact operates for a time equal to its rated electrical life. Ratings of mechanical life are for dry circuits (drawing no current).

Some contacts are also rated for an inductive load in addition to their resistive load rating. A third common rating is a motor or lamp rating for loads that draw much higher initial current than the normal steady state current.

All of these ratings assume that no contact protection is used. If proper contact protection is used, the rated voltage and/or current can be handled for a greater number of operations or a higher voltage and/or current can be handled for the rated number of operations.

LOADS WITH HIGH INRUSH CURRENTS

If the load is not resistive, the contacts must be appropriately derated or protected. Lamps, motors, and capacitive loads all draw much higher current when the contacts are closed than their steady-state current. The

---

*A small amount of arcing may actually be useful in burning off any thin insulating film that has formed on the contacts.

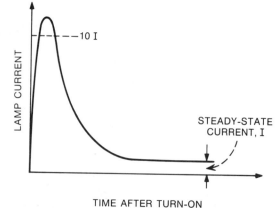

Figure 7-3. *Lamp current versus time.*

initial current in a lamp filament, for example, can be 10–15 times the normal rated current, as shown in Fig. 7-3. Typically, a contact is rated at only 20% of its normal resistive load capacity for lamp loads.

Capacitive loads also can draw extremely high initial currents. The charging current of a capacitor is limited only by the series resistance of the external circuit.

Motors typically draw initial currents that are 5–10 times their normal rated currents. In addition, the motor inductance causes a high voltage to be generated when the current is interrupted (an inductive kick). This also causes arcing. Motors, therefore, are difficult to switch since they cause contact damage on both "make" and "break."

To protect a contact used in a circuit with high inrush current, the initial current must be limited. Using a resistor in series with the contact to limit the initial current is not always feasible, since it also limits the steady-state current. If a resistor is not satisfactory, a low dc resistance inductor can be used to limit the initial current. In some light duty applications, ferrite beads placed on the contact lead may provide sufficient initial current limiting without affecting the steady-state current.

In severe cases, a switchable current-limiting resistor, as shown in Fig. 7-4, may be required. Here a relay is placed across the capacitive load with its normally open contact placed across the current limiting resistor. When the switch is closed, the capacitor charging current is limited by resistor R. When the voltage across the capacitor becomes large enough to operate the relay, the normally open contact closes, shorting out the current limiting resistor.

Another problem associated with closing contacts is chatter, or bounce.

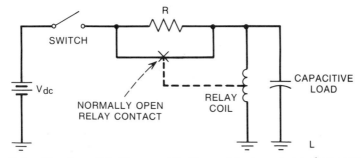

Figure 7-4.   *Use of a switchable current limiting resistor to protect a closing contact.*

After the contacts initially touch, they may bounce open again and break the circuit. In some contacts, this may continue for ten or more times, and each time the contacts must make and break the current. Not only can the repeated arcing cause operational problems in the circuit, but it produces considerably more contact damage and high frequency radiation.

INDUCTIVE LOADS

The voltage across an inductance (L) is given by

$$V = L\left(\frac{di}{dt}\right). \tag{7-1}$$

This expression explains the large voltage transient encountered when the current through an inductor is suddenly interrupted. The rate of change, di/dt, becomes large and negative, resulting in the large reverse voltage transient or inductive "kick." Theoretically, if the current goes from some finite value to zero instantaneously, the induced voltage would be infinite. But in reality, contact arcing and circuit capacitance never let this happen. Nevertheless, very large induced voltages do occur. Suppression of high voltage inductive transients consists of minimizing the di/dt term.

It is not at all uncommon for an inductance operating from a 26-V dc power supply to generate voltages of 500–5000 V when the current through the inductor is suddenly interrupted. Figure 7-5 shows the voltage waveshape across an inductor as the current is interrupted. The high voltage generated when a contact breaks the current to an inductive load causes severe contact damage. It is also the source of radiated and conducted noise, unless appropriate contact protection circuits are used. Under such conditions most of the energy stored in the inductance is dissipated in the arc, causing excessive damage.

The circuit shown in Fig. 7-6 can be used to illustrate the damage done to a set of contacts by an inductive load. In this figure, a battery is

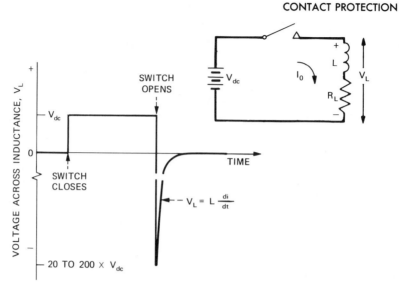

Figure 7-5. *Voltage across inductive load when switch closes and opens.*

connected to an inductive load through a switch contact. The load is assumed to have negligible resistance. In practice, this condition can be approximated by a low resistance dc motor. The steady state current becomes limited by the back EMF of the motor rather than the circuit resistance. Then let the switch be opened while a current I_0 is flowing through the inductance. The energy stored in the magnetic field of the inductance is equal to $(1/2)LI_0^2$. When the switch is opened, what happens to the energy stored in the magnetic field of the inductance? If the circuit resistance is negligible, all of the energy must be dissipated in an arc that forms across the contacts or be radiated. Without some type of protective circuitry, a switch used in this application does not last very long.

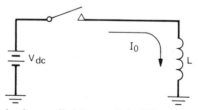

Figure 7-6. *Inductive load controlled by a switch. When the switch opens, most of the energy stored in the inductor is dissipated in the arc formed across the switch contacts.*

CONTACT PROTECTION FUNDAMENTALS

Figure 7-7 summarizes the conditions for contact breakdown in terms of the required voltage-distance relationships. The required breakdown voltage for starting a glow discharge is shown, as is the minimum voltage required to sustain the glow discharge. Also shown is a voltage gradient of 0.5 MV/cm, which is that required to produce an arc discharge. The minimum voltage required to maintain the arc discharge is also shown in this figure. The heavy line, therefore, represents the composite requirements for producing contact breakdown. To the right and below this curve there is no breakdown, whereas above and to the left of this curve, contact breakdown occurs.

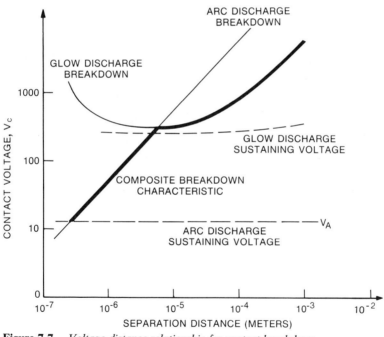

Figure 7-7. *Voltage-distance relationship for contact breakdown.*

A more useful presentation of the breakdown information contained in Fig. 7-7 is to plot the breakdown voltage versus time, instead of distance. This conversion can be accomplished using the separation velocity of the contacts. A typical composite breakdown characteristic as a function of time is shown in Fig. 7-8. As can be seen, there are two requirements for avoiding contact breakdown:

1. *Keep the contact voltage below 300 V to prevent a glow discharge.*

2. *Keep the initial rate of rise of contact voltage below the value necessary to produce an arc discharge. (A value of 1 V/μs is satisfactory for most contacts.)*

If it is not possible to avoid contact breakdown in a specific application, the breakdown should be kept from being self-sustaining. This usually means arranging the circuit so that the current available is always below that necessary to sustain the breakdown.

To determine whether or not breakdown can occur in a specific case, it is necessary to know what voltage is produced across the contacts as they open. This voltage is then compared to the breakdown characteristics in Fig. 7-8. If the contact voltage is above the breakdown characteristic, contact breakdown occurs.

Figure 7-9 shows an inductive load connected to a battery through a switch S. The voltage that would be produced across the contacts of the opening switch if no breakdown occurred, is called the "available circuit voltage." This is shown in Fig. 7-10 for the circuit in Fig. 7-9. I_0 is the current flowing through the inductor the instant the switch is opened, and C is the stray capacitance of the wiring. Figure 7-11 compares the available circuit voltage (Fig. 7-10) to the contact breakdown characteristics (Fig. 7-8). The voltage exceeds the breakdown characteristics from t_1 to t_2 and, therefore, contact breakdown occurs in this region.

Knowing that breakdown occurs, let us consider in more detail exactly what happens as the contacts in Fig. 7-9 are opened. When the switch is

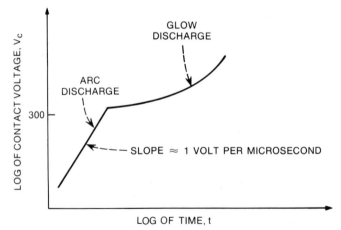

Figure 7-8. *Contact breakdown characteristics versus time.*

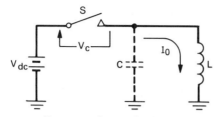

Figure 7-9. *Contact controlling an inductive load. Capacitor C represents the stray capacitance of the wiring.*

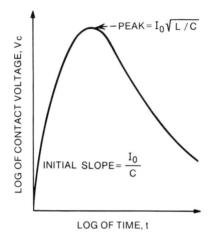

Figure 7-10. *Available circuit voltage across opening contact for circuit in Fig. 7-9, assuming no contact breakdown.*

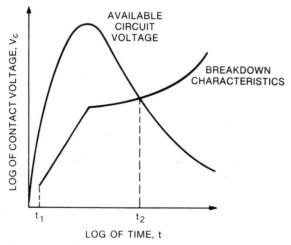

Figure 7-11. *Comparison of available circuit voltage and contact breakdown characteristics for circuit in Fig. 7-9.*

183

opened, the magnetic field of the inductance tends to keep the current I_0 flowing. Since the current cannot flow through the switch, it flows through the stray capacitance C instead. This charges the capacitor and the voltage across the capacitor rises at an initial rate of I_0/C, as shown in Fig. 7-12. As soon as this voltage exceeds the breakdown curve, an arc occurs across the contacts. If the available current at this point is less than the minimum arcing current I_A, the arc lasts only long enough to discharge the capacitance C to a voltage below the sustaining voltage V_A. After the capacitor is discharged, the current again charges C and the process is repeated until the voltage exceeds the glow discharge voltage (point A in Fig. 7-12). At this point a glow discharge occurs. If the smaller sustaining current, necessary to maintain the glow discharge is still not available, the glow lasts only until the voltage drops below the minimum glow voltage V_G.* This process is repeated until time t_1, after which sufficient voltage is not available to produce any additional breakdowns.

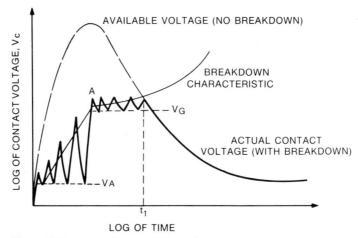

Figure 7-12. *Actual contact voltage for circuit in Fig. 7-9.*

If at any time the available current exceeds the minimum arcing current I_A, a steady arc occurs and continues until the available voltage or current falls below the minimum glow voltage or current. Figure 7-13 shows the waveshape when sufficient current is available to maintain a glow discharge, but not enough for an arc discharge.

If the stray capacitance C is increased sufficiently, or if a discrete

---

*If sufficient current is now available the glow discharge may transfer to an arc, and the voltage falls to V_A instead of V_G. Sufficient current however, is usually not available at the low voltage V_A to maintain the arc, so it is extinguished at this point.

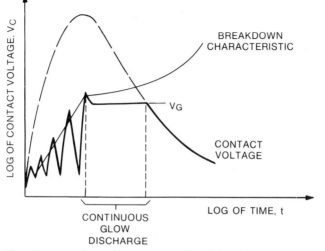

Figure 7-13. *Contact voltage for circuit in Fig. 7-9 when current is sufficient to maintain a continuous glow discharge.*

capacitor is placed in parallel with it, the peak voltage and the initial rate of rise of contact voltage can be reduced to the point where no arcing occurs. This waveshape is shown in Fig. 7-14. Using a capacitor this way, however, causes contact damage on closure due to the large capacitor charging current.

The electrical oscillations that occur in the resonant circuit of Fig. 7-9,

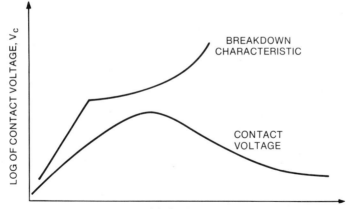

Figure 7-14. *Contact voltage for circuit in Fig. 7-9 when capacitance is large enough to prevent breakdown.*

when the switch is opened, can become the source of high-frequency interference to nearby equipment. These oscillations can be avoided if sufficient resistance and capacitance are provided to guarantee that the circuit is overdamped. The required condition for nonoscillation is given in the section on contact protection networks for inductive loads on page 192.

TRANSIENT SUPPRESSION FOR INDUCTIVE LOADS

To protect contacts that control inductive loads, and to minimize radiated and conducted noise, some type of contact protection network must normally be placed across the inductance, the contacts, or both. In some cases the protection network can be connected across either the load or the contact with equal effectiveness. In large systems a load may be controlled by more than one contact, and it may be more economical to provide protection at the load rather than at each individual contact.

In severe cases protection networks may have to be applied across both the inductance and the contacts to eliminate interference and protect the contacts adequately. In other cases the amount of protection that can be used is limited by operational requirements. For example, protection networks across the coil of a relay increase the release time. In this case the protection network has to be a compromise between meeting operational requirements and providing adequate protection to the contacts controlling the relay.

From a noise reduction point of view it is usually preferable to provide as much transient suppression as possible across the noise source—in this case, the inductor. In most cases this provides sufficient protection for the contacts. When it is not, additional protection can be used across the contacts.

Precise calculations for the component values of a contact protection network are difficult. It involves parameters, the values of which are normally unknown by the circuit designer, such as the inductance and capacitance of the interconnecting wiring and the contact separation velocity. The simplified design equations that follow are a starting point, and in many cases, provide an acceptable contact protection network. Empirical tests should be used, however, to verify the effectiveness of the network in the intended application.

Protection networks can be divided into two categories, those that are usually applied across the inductor, and those that are usually applied across the contacts. Some of these networks, however, can be applied in either place.

Figure 7-15 shows six networks commonly placed across a relay coil or other inductance to minimize the transient voltage generated when current is interrupted. In Fig. 7-15A a resistor is connected across the inductor.

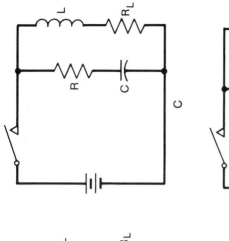

A

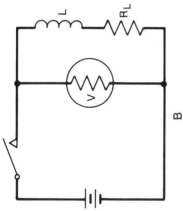

B

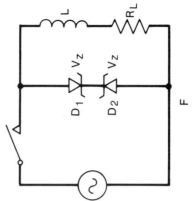

C

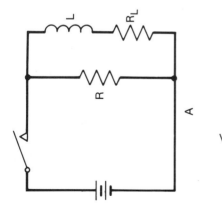

D

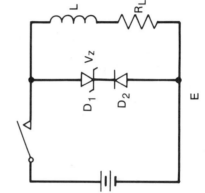

E

F

Figure 7-15. *Networks used across load to minimize "inductive kick" produced by an inductor when the current is interrupted.*

187

When the switch opens, the inductor drives whatever current was flowing before the opening of the contact through the resistor. The transient voltage peak therefore increases with increasing resistance, but is limited to the steady-state current times the resistance. If R is made equal to the load resistance R_L, the voltage transient is limited to a magnitude equal to the supply voltage. In this case the voltage across the contact is the supply voltage plus the induced coil voltage, or twice the supply voltage. This circuit is very wasteful of power since the resistor draws current whenever the load is energized. If R should equal the load resistance, the resistor dissipates as much steady-state power as the load.

Another arrangement is shown in Fig. 7-15B, where a varistor (a voltage variable resistor) is connected across the inductor. When the voltage across the varistor is low, its resistance is high, but when the voltage across it is high its resistance is low. This device works the same as the resistor in Fig. 7-15A, except that the power dissipated by the varistor while the circuit is energized is reduced.

A better arrangement is shown in Fig. 7-15C, here a resistor and capacitor are connected in series and placed across the inductor. This circuit dissipates no power when the inductor is energized. When the contact is opened, the capacitor initially acts as a short circuit, and the inductor drives its current through the resistor. The values for the resistor and the capacitor can be determined by the method described on page 191 for the R–C network.

In Fig. 7-15D a semiconductor diode is connected across the inductor. The diode is poled so that when the circuit is energized, no current flows through the diode. However, when the contact opens, the voltage across the inductor is of opposite polarity than that caused by the battery. This voltage forward biases the diode, which then limits the transient voltage across the inductor to a very low value (the forward voltage drop of the diode plus any IR drop in the diode). The voltage across the opening contact is therefore approximately equal to the supply voltage. This circuit is very effective in suppressing the voltage transient. However, the time required for the inductor current to decay is more than for any of the previous circuits and may cause operational problems.

For example, if the inductor is a relay, its release time is increased. A small resistor can be connected in series with the diode in Fig. 7-15D to decrease the release time of the relay, but at the expense of generating a higher transient voltage. The diode must have a voltage rating greater than the maximum supply voltage. The current rating of the diode must be greater than the maximum load current. If the contacts only operate occasionally, the peak current rating of the diode can be used. If the contacts operate more than a few times per minute, the continuous current rating of the diode should be used.

Adding a zener diode in series with a rectifier diode, as shown in Fig. 7-15E, allows the inductor current to decay faster. This protection, however, is not as good as that for the diode above and uses an extra component. In this case, the voltage across the opening contact is equal to the zener voltage plus the supply voltage.

Neither of the diode circuits (Fig. 7-15D or E) can be used with ac circuits. Circuits that operate from ac sources or circuits that must operate from two dc polarities can be protected using the networks in Fig. 7-15A through C, or by two zener diodes connected back to back, as shown in Fig. 7-15F. Each zener must have a voltage breakdown rating greater than the peak value of the ac supply voltage and a current rating equal to the maximum load current.

CONTACT PROTECTION NETWORKS FOR INDUCTIVE LOADS

C-Network

Figure 7-16 shows three contact protection networks commonly used across contacts that control inductive loads. One of the simplest methods of suppressing arcs due to interrupting dc current is to place a capacitor across the contact, as shown in Fig. 7-16A. If the capacitor is large enough, the load current is momentarily diverted through it as the contact is opened, and arcing does not occur. However, when the contact is open the capacitor charges up to the supply voltage V_{dc}. When the contact is then closed, the capacitor discharges through the contact with the initial discharge current limited only by the parasitic resistance of the wiring and the contacts.

The larger the value of the capacitor and the higher the supply voltage, the more damage the arc on "make" does, due to the increased energy stored in the capacitor. If the contacts bounce on "make" additional damage is done due to multiple making and breaking of the current. Because of these reasons, using a capacitor alone across a set of contacts is not generally recommended. If used, the value of capacitance is determined as explained in the following section.

R–C Network

Figure 7-16B shows a circuit that overcomes the disadvantage of the circuit in Fig. 7-16A by limiting the capacitor discharge current when the contact is closed. This is done by placing a resistor, R, in series with the capacitor. For contact closing, it is desirable to have the resistance as large as possible to limit the discharge current. However, when the contact is opened, it is desirable to have the resistance as small as possible, since the resistor decreases the effectiveness of the capacitor in preventing arcing.

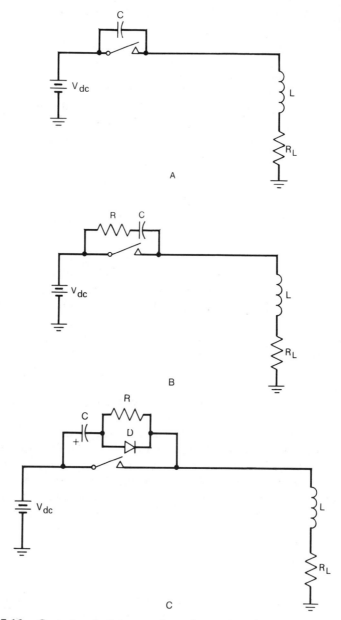

Figure 7-16. *Contact protection networks used across switch contacts.*

The actual value of R must therefore be a compromise between the two conflicting requirements.

The minimum value of R is determined by closing conditions. It can be set by limiting the capacitor discharge current to the minimum arcing current $I_A{}^*$ for the contact. The maximum value is determined by opening conditions. The initial voltage across the opening contact is equal to I_0R. If R is equal to the load resistance, the instantaneous voltage across the contact equals the supply voltage. The maximum value of R is usually taken equal to the load resistance to limit the initial voltage developed across the opening contacts to the supply voltage. The limits on R can, therefore, be stated as

$$\frac{V_{dc}}{I_A} < R < R_L,\tag{7-2}$$

where R_L is equal to the load resistance.

The value of C is chosen to meet two requirements: (1) the peak voltage across the contacts should not exceed 300 V (to avoid a glow discharge), and (2) the initial rate of rise of contact voltage should not exceed 1 V per μs (to avoid an arc discharge). The latter requirement is satisfied if C is at least 1 μF/A of load current.

The peak voltage across the capacitor is usually calculated by neglecting the circuit resistance and assuming all the energy stored in the inductive load is transferred to the capacitor. Under these conditions

$$V_{C(\text{peak})} = I_0\sqrt{L/C} ,\tag{7-3}$$

where I_0 is the current through the load inductance when the contact is opened. The value of the capacitor C should always be chosen so that V_C (peak) does not exceed 300 V. Therefore

$$C \geq \left(\frac{I_0}{300}\right)^2 L.\tag{7-4}$$

In addition, to limit the initial rate of rise of contact voltage to 1 volt per microsecond.

$$C \geq I_0 \times 10^{-6}\tag{7-5}$$

In some cases it is preferable that the resonant circuit formed by the

---

*Limiting the discharge current to $0.1I_A$ is preferable. However, since the value of the resistor R is a compromise between two conflicting requirements this usually cannot be done in the case of the R–C network.

inductor and capacitor be nonoscillating (overdamped). The condition for nonoscillation is

$$C \geq \frac{4L}{R_1^2},\tag{7-6}$$

where R_1 is the total resistance in series with the L-C circuit. In the case of Fig. 7-16B this would be $R_1 = R_L + R$. The requirement for nonoscillation, however, is not usually adhered to since it requires a large value capacitor.

The R–C protection network is the most widely used because of its low cost and small size. In addition, it only has a small effect on the release time of the load. The R–C network is not, however, 100% effective. The presence of the resistor causes an instantaneous voltage (equal to I_0R) to develop across the opening contact and therefore, some early arcing is present. Figure 7-17 shows the voltage developed across the contact, with a properly designed R–C network, superimposed on the contact breakdown characteristic. This figure shows the early arcing due to the instantaneous voltage increase across the contact.

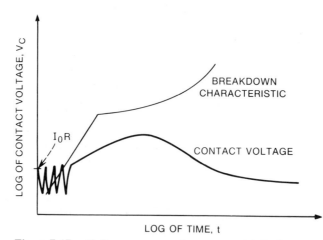

Figure 7-17.   *Voltage across opening contact with R–C protection network.*

R-C-D Network

Figure 7-16C shows a more expensive circuit that overcomes the disadvantages of the circuits in Fig. 7-16A and B. When the contact is open, capacitor C charges up to the supply voltage with the polarity shown in the figure. When the contact closes, the capacitor discharges through resistor R, which limits the current. When the contact opens, however, diode D

shorts out the resistor, thus allowing the load current to momentarily flow through the capacitor while the contact opens. The diode must have a breakdown voltage greater than the supply voltage with an adequate surge current rating (greater than the maximum load current). The capacitor value is chosen the same as for the R–C network. Since the diode shorts out the resistor when the contacts open, a compromise resistance value is no longer required. The resistance can now be chosen to limit the current on closure to less than one tenth the arcing current,

$$R \geq \frac{10 V_{dc}}{I_A}. \tag{7-7}$$

The R–C–D network provides optimum contact protection, but it is more expensive than other methods and cannot be used in an ac circuit.

INDUCTIVE LOADS CONTROLLED BY A TRANSISTOR SWITCH

If an inductive load is controlled by a transistor switch, care must be taken to guarantee that the transient voltage generated by the inductor when the current is interrupted does not exceed the breakdown voltage of the transistor. One of the most effective, and common, ways to do this is to place a diode across the inductor, as shown in Fig. 7-18. In this circuit the diode clamps the transistor collector to $+V$ when the transistor interrupts the current through the inductor, thereby limiting the voltage across the transistor to $+V$. Any of the networks of Fig. 7-15 may also be used. A zener diode connected across the transistor is another common method. In any case the network should be designed to limit the voltage across the transistor to less than its breakdown voltage rating.

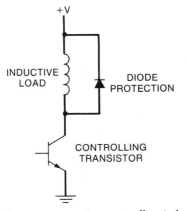

Figure 7-18.   *Diode used to protect transistor controlling inductive loads.*

RESISTIVE LOAD CONTACT PROTECTION

In the case of resistive loads operating with a source voltage of less than 300 V, a glow discharge cannot be started and, therefore, is of no concern. If the supply voltage is greater than the minimum arcing voltage V_A (about 12 V), an arc discharge occurs when the contacts are either opened or closed. Whether the arc, once started, sustains itself, depends on the magnitude of the load current.

If the load current is below the minimum arcing current I_A, the arc is quickly extinguished after initially forming. In this case, only a minimal amount of contact damage occurs, and in general, no contact protection networks are needed. Due to parasitic circuit capacitance or contact bounce, the arc starts, stops, and reignites many times. This type of arcing may be the source of high frequency radiation and may require some protection to control interference.

If the load current is greater than the minimum arcing current I_A a steady arc forms. This steady arc does considerable damage to the contacts. If the current is less than the resistive circuit current rating of the contact however, and the rated number of operations is satisfactory, contact protection may not be required.

If contact protection is required for a resistive load, what type of network should be used? In a resistive circuit the maximum voltage across an opening or closing contact is the supply voltage. Therefore, provided the supply voltage is under 300 V, the contact protection network does not have to provide high voltage protection. This function is already provided by the circuit. The required function of the contact protection network, in this case, is to limit the initial rate of rise of contact voltage to prevent initiating an arc discharge. This can best be accomplished by using the $R–C–D$ network in Fig. 7-16C across the contact.

CONTACT PROTECTION SELECTION GUIDE

The following guide can be used to determine the type of contact protection for various loads:

1. *Noninductive loads drawing less than the arcing current*, in general, require no contact protection.

2. *Inductive loads drawing less than the arcing current* should have an $R–C$ network or a diode for protection.

3. *Inductive loads drawing greater than the arcing current* should have an $R–C–D$ network or a diode for protection.

4. *Noninductive loads drawing greater than the arcing current* should use the $R–C–D$ network. Equation 7-4 does not have to be satisfied in this case provided the supply voltage is less than 300 V.

EXAMPLES

Proper selection of contact protection may be better understood with some numerical examples.

Example 7-1. A 150 Ω, 0.2 H relay coil is operated from a 12-V dc power source through a silver switch contact. The problem is to design a contact protection network for use across the relay.

The steady state load current is 80 mA, which is less than the arcing current for silver contacts; therefore, an *R–C* network or diode is appropriate. To keep the voltage gradient across the contact below 1 V/μs, the capacitance of the protection network must be greater than 0.08 μF (from Eq. 7-5). To keep the maximum voltage across the opening contact below 300 V, the capacitance must be greater than 0.014 μF (from Eq. 7-4). From Eq. 7-2, the value of the resistor should be between 30 and 150 Ω. Therefore, an appropriate contact protection network is 0.1 μF in series with 100 Ω placed either across the contact or load.

Example 7-2. A magnetic brake having 1 H inductance and 53 Ω resistance is operated from a 48-V dc source through a switch with silver contacts. If an *R–C* contact protection network is used, the resistor should have a value (from Eq. 7-2) of 120 < *R* < 53. Since this is impossible, a more complicated protection network must be used, such as the *R–C–D* network. For the *R–C–D* network, the resistor should have a value greater than 1200 Ω (from Eq. 7-7). The steady state dc current in the brake is 0.9 A. Therefore, from Eq. 7-5, the capacitor must be greater than 0.9 μF to limit the voltage gradient across the contacts on opening. From Eq. 7-4, the capacitor must also be greater than 9 μF. A 10-μF capacitor with a 300-V rating could be used, with a 1500 Ω resistor and a diode, as shown in Fig. 7-19.

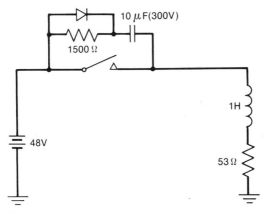

Figure 7-19. *Contact protection network for Example 7-2.*

The 10-μF, 300-V capacitor must of necessity be relatively large physically. To avoid using such a large capacitor, the following alternate solution could be used. If a series combination of a 60-V zener diode and a rectifier diode is placed across the load, the maximum transient voltage across the load would be limited to 60 V. The maximum voltage across the contact upon opening would then be the zener voltage plus the supply voltage, or 108 V. Therefore, the capacitor in the protection network does not have to be chosen to limit the maximum voltage across the contacts to 300 V, since this voltage is already limited by the diode to 108 V. The only requirement now on the capacitor is that it satisfy Eq. 7-5. Therefore, a 1-μF, 150-V capacitor can be used as shown in Fig. 7-20, which avoids the need for a physically large size, 10-μF, 300-V capacitor.

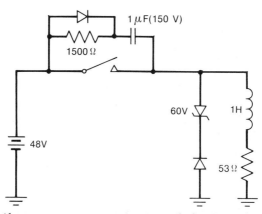

Figure 7-20. *Alternate contact protection network for Example 7-2. This circuit allows the use of a physically smaller capacitor.*

SUMMARY

- Two types of breakdown are important in switching contact: the glow, or gas discharge, and the arc, or metal-vapor, discharge.
- To prevent a glow discharge, keep contact voltage below 300 V.
- To prevent an arc discharge, keep the initial rate of rise of contact voltage to less than 1 V/μs.
- Lamps and capacitor loads cause contact damage on closure due to the high inrush currents.
- Inductive loads are most damaging due to the high voltages they generate when current is interrupted.
- The R–C network is the most widely used protection network.
- The R–C–D network or the diode are the most effective protection network.

- The effect of the contact protection network on the release time of inductive loads must be considered.
- A diode connected across an inductor is a very effective transient suppression network, however, it may cause operational problems since it prevents the rapid decay of the inductor current.

BIBLIOGRAPHY

Auger, R. W., and Puerschner, K., *The Relay Guide*, Reinhold, New York, 1960.

Bell Laboratories, *Physical Design of Electronic Systems* Vol. III, Integrated Device and Connection Technology, Chapter 9 (Performance Principles of Switching Contacts), Prentice-Hall, Englewood Cliffs, N. J., 1971.

Dewey, R., "Everyone Knows That Inductive Loads Can Greatly Shorten Contact Life," *EDN*, April 5, 1973.

Duell, J. P. Jr., "Get Better Price/Performance From Electrical Contacts", *EDN*, June 5, 1973.

Holm, R., *Electrical Contacts*, Fourth Edition, Springer-Verlag, Berlin, 1967.

National Association of Relay Manufacturers, *Engineers' Relay Handbook*, Second Edition, Hayden Book Co., New York, 1969.

Oliver, F. J., *Practical Relay Circuits*, Hayden Book Co., New York, 1971.

8 INTRINSIC NOISE SOURCES

Even if all external noise coupling could be eliminated from a circuit, a theoretical minimum noise level would still exist due to certain intrinsic or internal noise sources. Although the rms value of these noise sources can be well defined, the instantaneous amplitude can only be predicted in terms of probability. Intrinsic noise is present in almost all electronic components.

This chapter covers the three most important intrinsic noise sources—thermal noise, shot noise, and contact noise. In addition, popcorn noise and methods of measuring random noise are discussed.

THERMAL NOISE

Thermal noise comes from thermal agitation of electrons within a resistance, and it sets a lower limit on the noise present in a circuit. Thermal noise is also referred to as resistance noise or "Johnson noise" (for J. B. Johnson, its discoverer). Johnson (1928) found that a nonperiodic voltage exists in all conductors and its magnitude is related to temperature. Nyquist (1928) subsequently described the noise voltage mathematically, using thermodynamic reasoning. He showed that the open-circuit rms noise voltage produced by a resistance is

$$V_t = \sqrt{4kTBR} \, , \tag{8-1}$$

where

k = Boltzmann's constant $(1.38 \times 10^{-23}$ joules$/°$K$)$,

T = Absolute temperature $(°$K$)$,

B = Noise bandwidth (Hz),

R = Resistance (Ω).

At room temperature $(290°$K$)$, $4kT$ equals 1.6×10^{-20} W/Hz. The bandwidth B in Eq. 8-1 is the equivalent noise bandwidth of the system being considered. The calculation of equivalent noise bandwidth is covered on page 204.

198

Thermal noise is present in all elements containing resistance. A plot of the thermal noise voltage at a temperature of 17°C (290°K) is shown in Fig. 8-1. Normal temperature variations have a small effect on the value of the thermal noise voltage. For example, at 117°C the noise voltage is only 16% greater than that given in Fig. 8.1 for 17°C.

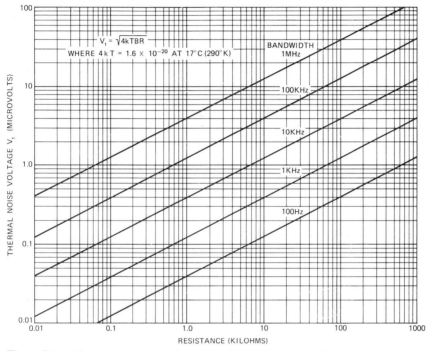

Figure 8-1. *Thermal noise voltage as a function of resistance and bandwidth.*

Equation 8-1 shows that the thermal noise voltage is proportional to the square root of the bandwidth and the square root of resistance. It would, therefore, be advantageous to minimize the resistance and bandwidth of a system to reduce the thermal noise voltage. If thermal noise is still a problem, considerable reduction is possible by operating the circuit at extremely low temperatures (close to absolute zero), or by using a parametric amplifier. Since the gain of a parametric amplifier comes from a reactance varied at a rapid rate, it does not have thermal noise.

The thermal noise in a resistor can be represented by adding a thermal noise voltage source V_t in series with the resistor, as shown in Fig. 8-2. The magnitude of V_t is determined from Eq. 8-1. In some cases it is preferable to represent the thermal noise by an equivalent rms noise current generator of magnitude

$$I_t = \sqrt{\frac{4kTB}{R}} \qquad (8\text{-}2)$$

in parallel with the resistor. This is also shown in Fig. 8-2.

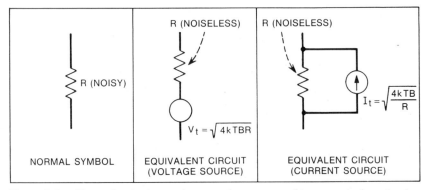

Figure 8-2. *Thermal noise in a resistor can be represented in an equivalent circuit as a voltage source (center) or a current source (right).*

Thermal noise is a universal function, independent of the composition of the resistance. For example, a 1000-Ω carbon resistor has the same amount of thermal noise as a 1000-Ω tantalum thin-film resistor. An actual resistor may have more noise than that due to thermal noise, but never less. This additional, or excess, noise is due to the presence of other noise sources. A discussion of noise in actual resistors was given in Chapter 5.

Electric circuit elements can produce thermal noise only if they are capable of dissipating energy. Therefore, a reactance cannot produce thermal noise. This can be demonstrated by considering the example of a resistor and capacitor connected, as shown in Fig. 8-3. Here we make the erroneous assumption that the capacitor generates a thermal noise voltage V_{tc}. The power that generator V_{tc} delivers to the resistor is $P_{cr} = N(f)V_{tc}^2$, where $N(f)$ is some nonzero network function.* The power that generator V_{tr} delivers to the capacitor is zero, since the capacitor cannot dissipate power. For thermodynamic equilibrium, the power that the resistor delivers to the capacitor must equal the power that the capacitor delivers to

---

*In this example $N(f) = \dfrac{1}{R}\left(\dfrac{j\omega}{j\omega + 1/RC}\right)^2$.

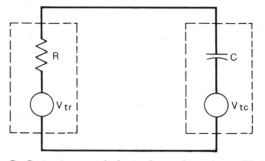

Figure 8-3. *An R–C circuit can only be in thermodynamic equilibrium if V_{tc} equals zero.*

the resistor. Otherwise, the temperature of one component increases and the temperature of the other component decreases. Therefore,

$$P_{cr} = N(f)V_{tc}^2 = 0. \qquad (8\text{-}3)$$

The function $N(f)$ cannot be zero at all frequencies because it is a function of the network. Voltage V_{tc} must therefore be zero, thus demonstrating that a capacitor cannot generate thermal noise.

Let us now connect two unequal resistors (at the same temperature) together as shown in Fig. 8-4, and check for thermodynamic equilibrium. The power that generator V_{t1} delivers to resistor R_2 is

$$P_{12} = \frac{R_2}{(R_1 + R_2)^2} V_{t1}^2. \qquad (8\text{-}4)$$

Substituting Eq. 8-1 for V_{t1} gives

$$P_{12} = \frac{4kTBR_1R_2}{(R_1 + R_2)^2}. \qquad (8\text{-}5)$$

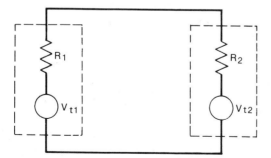

Figure 8-4. *Two resistors connected in parallel are in thermodynamic equilibrium.*

The power that generator V_{t2} delivers to R_1 is

$$P_{21} = \frac{R_1}{(R_1 + R_2)^2} V_{t2}^2.$$ (8-6)

Substituting Eq. 8-1 for V_{t2} gives

$$P_{21} = \frac{4kTBR_1R_2}{(R_1 + R_2)^2}.$$ (8-7)

Comparing Eq. 8-5 to Eq. 8-7 we conclude that

$$P_{12} = P_{21},$$ (8-8)

thus showing that the two resistors are in thermodynamic equilibrium.

The power that generator V_{t1} delivers to resistor R_1 does not have to be considered in the above calculation. This power comes from and is dissipated in resistor R_1. Thus it produces no net effect on the temperature of resistor R_1. Similarly, the power that generator V_{t2} delivers to resistor R_2 need not be considered.

Let us now consider the case when the two resistors in Fig. 8-4 are equal in value, and maximum power transfer occurs between the resistors. We can then write

$$P_{12} = P_{21} = P_n = \frac{V_t^2}{4R}.$$ (8-9)

Substituting Eq. 8-1 for V_t gives

$$P_n = kTB \quad \text{watts.}$$ (8-10)

The quantity kTB is referred to as the "available noise power." At room temperature (17°C) this noise power per hertz of bandwidth is 4×10^{-21} W, and is independent of the value of the resistance.

It can be shown (van der Ziel, 1954, p. 17) that the thermal noise generated by any arbitrary connection of passive elements is equal to the thermal noise that would be generated by a resistance equal to the real part of the equivalent network impedance. This fact is useful for calculating the thermal noise of a complex passive network.

CHARACTERISTICS OF THERMAL NOISE

The frequency distribution of thermal noise power is uniform. For a specified bandwidth anywhere in the spectrum, the available noise power is constant and independent of the resistance value. For example, the noise power in a 100-Hz band between 100 and 200 Hz is equal to the noise power in a 100-Hz band between 1,000,000 and 1,000,100 Hz. When viewed on a wideband oscilloscope, thermal noise appears as shown in Fig. 8-5. Such noise—with a uniform power distribution with respect to frequency—is called "white noise," implying that it is made up of many frequency components. Many noise sources other than thermal noise share this characteristic, and are similarly referred to as white noise.

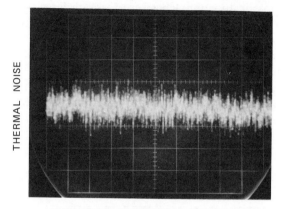

TIME, 200 MICROSECONDS PER DIVISION

Figure 8-5. *Thermal noise as seen on a wideband oscilloscope (horizontal sweep 200 μs per division).*

Although the rms value for thermal noise is well defined, the instantaneous value can only be defined in terms of probability. The instantaneous amplitude of thermal noise has a Gaussian, or normal, distribution. The average value is zero and the rms value is given by Eq. 8-1. Figure 8-6 shows the probability density function for thermal noise. The

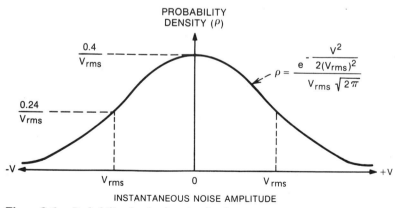

Figure 8-6. *Probability density function for thermal noise (Gaussian distribution).*

probability of obtaining an instantaneous voltage between any two values is equal to the integral of the probability density function between the two values. The probability density function is greatest at zero magnitude, indicating that values near zero are most common.

The crest factor of a waveform is defined as the ratio of the peak to the rms value. For thermal noise the probability density function, shown in Fig. 8-6, asymptotically approaches zero for both large positive and large negative amplitudes. Since the curve never reaches zero, there is no finite limit to the magnitude of the instantaneous noise voltage. On this basis the crest factor would be infinite, which is not a very useful result. A more useful result is obtained if we calculate the crest factor for peaks that occur at least a specified percentage of the time. Table 8-1 shows the results. Normally, only peaks that occur at least 0.01% of the time are considered, and *a crest factor of approximately 4 is used for thermal noise.*

Table 8-1   Crest Factors for Thermal Noise

Percent of time peak is exceeded	Crest factor (peak/rms)
1.0	2.6
0.1	3.3
0.01	3.9
0.001	4.4
0.0001	4.9

EQUIVALENT NOISE BANDWIDTH

The noise bandwidth B is the voltage-gain-squared bandwidth of the system or circuit being considered. The noise bandwidth is defined for a system with uniform gain throughout the passband and zero gain outside the passband. Fig. 8-7 shows this ideal response for a low pass circuit and a bandpass circuit.

Practical circuits do not have these ideal characteristics but have responses similar to those shown in Fig. 8-8. The problem, then, is to find an equivalent noise bandwidth that can be used in equations to give the same results as the actual non-ideal bandwidth does in practice. In the case of a white noise source (equal noise power for a specified bandwidth anywhere in the spectrum), the objective is met if the area under the equivalent noise bandwidth curve is made equal to the area under the actual curve. This is shown in Fig. 8-9 for a low pass circuit.

For any network transfer function, $A(f)$ (expressed as a voltage or

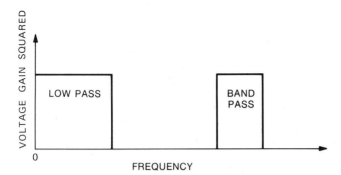

Figure 8-7. *Ideal bandwidth of low-pass and band-pass circuit elements.*

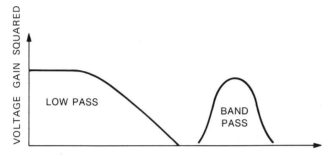

Figure 8-8. *Actual bandwidth of low-pass and band-pass circuit elements.*

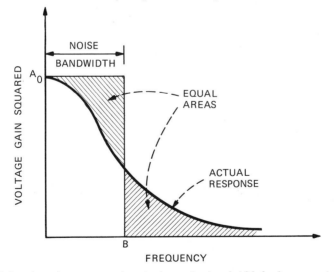

Figure 8-9. *Actual response and equivalent noise bandwidth for low-pass circuit. This curve is drawn with a linear scale.*

current ratio), there is an equivalent noise bandwidth with constant magnitude of transmission A_0 and bandwidth of

$$B = \frac{1}{|A_0|^2} \int_0^\infty |A(f)|^2 \, df. \tag{8-11}$$

A typical bandpass function is shown in Fig. 8-10. A_0 is usually taken as the maximum absolute value of $A(f)$.

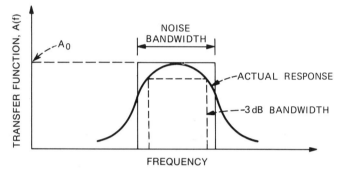

Figure 8-10. *Any network transfer function can be expressed as an equivalent bandwidth with constant transmission ratio.*

Example 8-1. Calculate the equivalent noise bandwidth for the simple $R{-}C$ circuit of Fig. 8-11. The voltage gain of this single pole (time constant) circuit versus frequency is

$$A(f) = \frac{f_0}{jf + f_0}, \tag{8-12}$$

where

$$f_0 = \frac{1}{2\pi RC}. \tag{8-13}$$

Frequency f_0 is where the voltage gain is down 3 dB, as shown in Fig. 8-11. At $f = 0$, $A(f) = A_0 = 1$. Substituting Eq. 8-12 into Eq. 8-11 gives

$$B = \int_0^\infty \left[\frac{f_0}{\sqrt{f_0^2 + f^2}} \right]^2 \, df, \tag{8-14a}$$

$$B = f_0^2 \int_0^\infty \left(f_0^2 + f^2 \right)^{-1} \, df. \tag{8-14b}$$

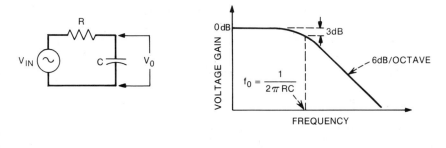

SCHEMATIC   TRANSFER FUNCTION

Figure 8-11. *R–C circuit.*

This can be integrated by letting $f = f_0 \tan \theta$, therefore, $df = f_0 \sec^2 \theta \, d\theta$. Making this substitution into Eq. 8-14b gives

$$B = f_0 \int_0^{\pi/2} d\theta. \tag{8-15}$$

Integrating gives

$$B = \frac{\pi}{2} f_0. \tag{8-16}$$

Therefore, the equivalent noise bandwidth for this circuit is $\pi/2$ or 1.57 times the 3-dB voltage bandwidth f_0. This result can be applied to any circuit that can be represented as a single-pole, low-pass filter. This result is also applicable to certain active devices, such as transistors, which can be modeled as single-pole, low-pass circuits.

Table 8-2 gives the ratio of the noise bandwidth to the 3-dB bandwidth for circuits with various numbers of identical poles. As can be seen, when the number of poles increase, the noise bandwidth approaches the 3-dB bandwidth. In the case of three or more poles the 3-dB bandwidth can be used in place of the noise bandwidth with only a small error.

Table 8-2   Ratio of the Noise Bandwidth B to the 3-dB Bandwidth f_0

Number of poles	B/f_0	High frequency rolloff (dB per octave)
1	1.57	6
2	1.22	12
3	1.15	18
4	1.13	24
5	1.11	30

A second method of determining noise bandwidth is to perform the integration graphically. This is done by plotting the voltage-gain-squared versus frequency on linear graph paper. A noise bandwidth rectangle is then drawn such that the area under the noise bandwidth curve is equal to the area under the actual curve, as shown in Fig. 8-9.

SHOT NOISE

Shot noise is associated with current flow across a potential barrier. It is due to the fluctuation of current around an average value resulting from the random emission of electrons (or holes). This noise is present in both vacuum tubes and semiconductors. In vacuum tubes, shot noise comes from the random emission of electrons from the cathode. In semiconductors, shot noise is due to random diffusion of carriers through the base of a transistor and the random generation and recombination of hole electron pairs.

The shot effect was analyzed theoretically by W. Schottky in 1918. He showed that the rms noise current was equal to (van der Ziel, 1954, p. 91)

$$I_{sh} = \sqrt{2qI_{dc}B} \ , \tag{8-17}$$

where

$$q = \text{Electron charge } (1.6 \times 10^{-19} \text{ coulombs}),$$

$$I_{dc} = \text{Average dc current (A)},$$

$$B = \text{Noise bandwidth (Hz)}.$$

Equation 8-17 is similar in form to Eq. 8-2. The power density for shot noise is constant with frequency and the amplitude has a Gaussian distribution. The noise is white noise and has the same characteristic as previously described for thermal noise. Dividing Eq. 8-17 by the square root of the bandwidth gives

$$\frac{I_{sh}}{\sqrt{B}} = \sqrt{2qI_{dc}} = 5.66 \times 10^{-10}\sqrt{I_{dc}} \ . \tag{8-18}$$

In Eq. 8-18 the noise current per square root of bandwidth is only a function of the dc current flowing through the device. Therefore, by measuring the dc current through the device, the amount of noise can be very accurately determined.

In making amplifier noise figure measurements (as discussed in Chapter 9), the availability of a variable source of white noise can considerably simplify the measurement. A diode can be used as a white noise source. If shot noise is the predominant noise source in the diode, the rms value of the noise current can be determined simply by measuring the dc current through the diode.

CONTACT NOISE

Contact noise is caused by fluctuating conductivity due to an imperfect contact between two materials. It occurs anywhere when two conductors are joined together, such as in switches and relay contacts. It also occurs in transistors and diodes due to imperfect contacts, and in composition resistors and carbon microphones that are composed of many small particles molded together.

Contact noise is also called by many other names. When found in resistors, it is referred to as "excess noise." When observed in vacuum tubes it is usually referred to as "flicker noise." Due to its unique frequency characteristic it is often called "$1/f$ noise," or "low frequency noise."

Contact noise is directly proportional to the value of direct current. The power density varies as the reciprocal of frequency ($1/f$) and the magnitude is Gaussian. The noise current I_f per square root of bandwidth can be expressed approximately (van der Ziel, 1954, p. 209) as

$$\frac{I_f}{\sqrt{B}} \approx \frac{KI_{dc}}{\sqrt{f}} , \qquad (8\text{-}19)$$

where

I_{dc} = Average value of dc current (A),

f = Frequency (Hz),

K = A constant that depends on the type of material and its geometry,

B = Bandwidth in hertz centered about the frequency (f).

It should be noted that the magnitude of contact noise can become very large at low frequencies due to its $1/f$ characteristic. Most of the theories advanced to account for contact noise predict that at some low frequencies the amplitude becomes constant. However, measurements of contact noise at frequencies as low as a few cycles per day still show the $1/f$ characteristic. Due to its frequency characteristics, contact noise is usually the most important noise source in low frequency circuits.

POPCORN NOISE

Popcorn noise, also called burst noise, was first discovered in semiconductor diodes and has recently reappeared in integrated circuits. If burst noise is amplified and fed into a loudspeaker, it sounds like corn popping, with thermal noise providing a background frying sound; thus the name pop-corn noise.

Unlike the other noise sources discussed in this chapter, popcorn noise is due to a manufacturing defect, and can be eliminated by improved manufacturing processes. This noise is caused by a defect in the junction, usually a metallic impurity, of a semiconductor device. Popcorn noise occurs in bursts and causes a discrete change in level, as shown in Fig. 8-12. The width of the noise bursts varies from microseconds to seconds. The repetition rate, which is not periodic, varies from several hundred pulses per second to less than one pulse per minute. For any particular sample of a device, however, the amplitude is fixed since it is a function of the characteristics of the junction defect. Typically, the amplitude is from 2 to 100 times the thermal noise.

The power density of popcorn noise has a $1/f^n$ characteristic, where n is typically two. Since the noise is a current-related phenomenon, popcorn noise voltage is greatest in a high impedance circuit, for example, the input circuit of an operational amplifier.

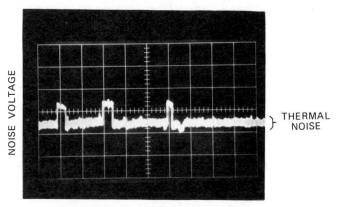

TIME, 20 MILLISECONDS PER DIVISION

Figure 8-12. *Output waveform of an IC op-amp with popcorn noise. The random noise on the baseline and at the top of the burst is thermal noise.*

ADDITION OF NOISE VOLTAGES

Noise voltages, or currents, produced independently with no relationships between each other are uncorrelated. When uncorrelated noise sources are added together, the total power is equal to the sum of the individual powers. Adding two noise voltage generators V_1 and V_2, together on a power basis, gives

$$V_{\text{total}}^2 = V_1^2 + V_2^2 \qquad (8\text{-}20)$$

The total noise voltage can then be written as

$$V_{total} = \sqrt{V_1^2 + V_2^2} \qquad (8-21)$$

Therefore, uncorrelated noise voltages can be added by taking the square root of the sum of the squares of the individual noise voltages.

Two correlated noise voltages can be added by using

$$V_{total} = \sqrt{V_1^2 + V_2^2 + 2\gamma V_1 V_2} \ , \qquad (8-22)$$

where γ is a correlation coefficient that can have any value from $+1$ to -1. When γ equals zero, the voltages are uncorrelated; when $|\gamma|$ equals 1, the voltages are totally correlated. For values of γ between 0 and $+1$ or 0 and -1 the voltages are partially correlated.

MEASURING RAMDOM NOISE

Noise measurements are usually made at the output of a circuit or amplifier. This is done for two reasons: (1) the output noise is larger and therefore easier to read on the meter, and (2) it avoids the possibility of the noise meter upsetting the shielding, grounding, or balancing of the input circuit of the device being measured. If a value of equivalent input noise is required, the output noise is measured and divided by the circuit gain to obtain the equivalent input noise.

Since most meters were intended to measure sinusoidal voltages, their response to a random noise source must be investigated. Three general requirements for a noise meter are: (1) it should respond to noise power, (2) it should have a crest factor of four or greater, and (3) its bandwidth should be at least 10 times the noise bandwidth of the circuit being measured. We will now consider the response of various types of meters when used to measure white noise.

A true rms meter is obviously the best choice, provided its bandwidth and crest factor are sufficient. A crest factor of three provides less than 1.5% error, whereas a crest factor of four gives an error of less than 0.5%. No correction to the meter indication is required.

The most common ac voltmeter responds to the average value of the waveform but has a scale calibrated to read rms. This meter uses a rectifier and a dc meter movement to respond to the average value of the waveform being measured. For a sine wave, the rms value is 1.11 times the average value. Therefore, the meter scale is calibrated to read 1.11 times the measured value. For white noise, however, the rms value is 1.25 times the average value. Therefore, when used to measure white noise, an average-responding meter reads too low. If the bandwidth and crest factor are sufficient, such a meter may be used to measure white noise by multiplying the meter reading by 1.13 or by adding 1.1 dB. Measurements should be made on the lower half of the meter scale to avoid clipping the peaks of the noise waveform.

Peak-responding voltmeters should not be used to measure noise since their response depends upon the charge and discharge time constants of the individual meter used.

An oscilloscope is an often overlooked, but excellent device for measuring white noise. One advantage it has over all other indicators is that the waveshape being measured can be seen. In this way you can be sure that you are measuring the desired random noise, not pickup or 60-Hz hum. The rms value of white noise is approximately equal to the peak-to-peak value taken from the oscilloscope, divided by eight.* When determining the peak-to-peak value on the oscilloscope, one or two peaks that are considerably greater than the rest of the waveform should be ignored. With a little experience, rms values can be accurately determined by this method. With an oscilloscope, random noise can be measured even when 60-Hz hum or other non-random noise sources are present, since the waveforms can be distinguished and measured separately on the display.

Table 8-3 summarizes the characteristics of various types of meters when used to measure white noise.

Table 8-3   Characteristics of Meters Used to Measure White Noise

Type of meter	Correction factor	Remarks
True rms	None	Meter bandwidth greater than ten times noise bandwidth, and meter crest factor three or greater.
RMS calibrated average responding	Multiply reading by 1.13 or add 1.1 dB.	Meter bandwidth greater than ten times noise bandwidth, and meter crest factor three or greater. Read below one-half scale to avoid clipping peaks.
RMS calibrated peak responding	Do not use	
Oscilloscope	$\text{RMS} \approx \frac{1}{8}$ peak-to-peak value	Waveshape can be observed to be sure it is random noise and not pickup. Ignore occasional, extreme peaks.

*This assumes a crest factor of four for white noise.

SUMMARY

- Thermal noise is present in all elements containing resistance.
- A reactance does not generate thermal noise.
- The thermal noise in any connection of passive elements is equal to the thermal noise that would be generated in a resistance equal to the real part of the equivalent network impedance.
- Shot noise is produced by current flow across a potential barrier.
- Contact noise ($1/f$ noise) is present whenever current flows through a non-homogeneous material.
- Contact noise is only a problem at low frequencies.
- Popcorn noise can be eliminated by improved manufacturing processes.
- The noise bandwidth is greater than the 3-dB bandwidth.
- As the number of poles (time constants) increase, the noise bandwidth approaches the 3-dB bandwidth.
- The crest factor for thermal noise is normally assumed to be four.
- Uncorrelated noise voltages add on a power basis, therefore

$$V_{total} = \sqrt{V_1^2 + V_2^2 + \cdots V_m^2} \ .$$

BIBLIOGRAPHY

Baxandall, P. J., "Noise in Transistor Circuits, Part 1," *Wireless World*, Vol. 74, November, 1968.

Bennett, W. R., "Characteristics and Origins of Noise—Part I," *Electronics*, Vol. 29, pp. 154–160, March, 1956.

Campbell, R. H., Jr., and Chipman, R. A., "Noise from Current-Carrying Resistors 20 to 500 Kc." *Proceedings of I.R.E.*, Vol. 37, pp. 938–942, August, 1949.

Dummer, G. W. A., *Fixed Resistors*, Sir Isaac Pitman, London, 1956.

Johnson, J. B., "Thermal Agitation of Electricity in Conductors," *Physical Review*, Vol. 32, pp. 97–109, July, 1928.

Lathi, B. P., *Signals, Systems and Communications*, Chapter 13, Wiley, New York 1965.

Mumford, W. W., and Scheibe, E. H., *Noise Performance Factors in Communication Systems*, Horizon House, Dedham, Mass., 1969.

Nyquist, H., "Thermal Agitation of Electric Charge in Conductors," *Physical Review*, Vol. 32, pp. 110–113, July, 1928.

Van der Ziel, A., *Fluctuation Phenomena in Semi-Conductors*, Academic, New York, 1959.

Van der Ziel, A., *Noise*, Prentice-Hall, Englewood Cliffs, N.J., 1954.

9 ACTIVE DEVICE NOISE

Bipolar transistors, field effect transistors (FETs), and integrated circuit operational amplifiers (op-amps) all have inherent noise generation mechanisms. This chapter discusses these internal noise sources and shows the conditions necessary to optimize noise performance.

Before covering active device noise, the general topics of how noise is specified and measured are presented. This general analysis provides a standard set of noise parameters that can then be used to analyze noise in various devices. The common methods of specifying device noise are: (1) noise factor, and (2) the use of a noise voltage and current model.

NOISE FACTOR

The concept of noise factor was developed in the 1940's as a method of evaluating noise in vacuum tubes. In spite of several serious limitations, the concept is still widely used today.

The noise factor (F) is a quantity that compares the noise performance of a device to that of an ideal (noiseless) device. It can be defined as

$$F = \frac{\text{Noise power output of actual device } (P_{no})}{\text{Noise power output of ideal device}} . \tag{9-1}$$

The noise power output of an ideal device is due to the thermal noise power of the source resistance. The standard temperature for measuring the source noise power is 290°K. Therefore, the noise factor can be written as

$$F = \frac{\text{Noise power output of actual device } (P_{no})}{\text{Power output due to source noise}} . \tag{9-2}$$

An equivalent definition of noise factor is the input signal-to-noise ratio divided by the output signal-to-noise ratio

$$F = \frac{S_i / N_i}{S_o / N_o} . \tag{9-3}$$

These signal-to-noise ratios must be power ratios unless the input imped-

214

ance is equal to the load impedance, in which case they can be voltage squared, current squared, or power ratios.

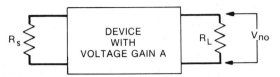

Figure 9-1. *Resistive source is used for noise factor measurements.*

All noise factor measurements must be taken with a resistive source, as shown in Fig. 9-1. The open circuit input noise voltage is, therefore, just the thermal noise of the source resistance R_s, or

$$V_t = \sqrt{4kTBR_s} \; . \qquad (9\text{-}4)$$

At 290°K, this is

$$V_t = \sqrt{1.6 \times 10^{-20} BR_s} \; . \qquad (9\text{-}5)$$

If the device has a voltage gain A, defined as the ratio of the output voltage measured across R_L to the open circuit source voltage, then the component of output voltage due to the thermal noise in R_s is AV_t. Using V_{no} for the total output noise voltage measured across R_L, the noise factor can be written as

$$F = \frac{(V_{no})^2/R_L}{(AV_t)^2/R_L} , \qquad (9\text{-}6)$$

or

$$F = \frac{(V_{no})^2}{(AV_t)^2} \; . \qquad (9\text{-}7)$$

V_{no} includes the effects of both the source noise and the device noise. Substituting Eq. 9-4 into Eq. 9-7 gives

$$F = \frac{(V_{no})^2}{4kTBR_s A^2} \; . \qquad (9\text{-}8)$$

The following three characteristics of noise factor can be seen by examining Eq. 9-8:

1. It is independent of load resistance R_L.

2. It does depend on source resistance R_s.

3. If a device were completely noiseless, the noise factor would equal one.

Noise factor expressed in decibels is called noise figure (*NF*) and is equal to

$$NF = 10 \log F. \qquad (9\text{-}9)$$

In a qualitative sense, noise figure and noise factor are the same, and in casual conversation they are often interchanged.

Due to the bandwidth term in the denominator of Eq. 9-8, there are two ways to specify the noise factor: (1) a spot noise, measured at a specified frequency over a 1-Hz bandwidth, or (2) an integrated, or average noise, measured over a specified bandwidth. If the device noise is "white" and is generated prior to the bandwidth-limiting portion of the circuit both the spot and integrated noise factors are equal. This is because, as the bandwidth is increased, both the total noise and the source noise increase by the same factor.

The concept of noise factor has three major limitations:

1. Increasing the source resistance may decrease the noise factor while increasing the total noise in the circuit.

2. If a purely reactive source is used, noise factor is meaningless since the source noise is zero, making the noise factor infinite.

3. When the device noise is only a small percentage of the source thermal noise (as with some low noise FETs), the noise factor requires taking the ratio of two almost equal numbers. This can produce inaccurate results.

A direct comparison of two noise factors is only meaningful if both are measured at the same source resistance. Noise factor varies with the bias conditions, frequency, and temperature as well as source resistance, and all of these should be defined when specifying noise factor.

Knowing the noise factor for one value of source resistance does not allow the calculation of the noise factor at other values of resistance. This is because both the source noise and device noise vary as the source resistance is changed.

MEASUREMENT OF NOISE FACTOR

A better understanding of noise factor can be obtained by describing the methods used to measure it. Two methods follow: (1) the single-frequency method, and (2) the noise-diode, or white noise, method.

The test set up for the single-frequency method is shown in Fig. 9-2. V_s is an oscillator set to the frequency of the measurement, and R_s is the source resistance. With the source V_s turned off, the output rms noise voltage V_{no} is measured. This voltage consists of two parts: the first due to the thermal noise voltage V_t of the source resistor, and the second due to the noise in the device.

$$V_{no} = \sqrt{(AV_t)^2 + (\text{Device Noise})^2} \ . \tag{9-10}$$

Figure 9-2. *Single frequency method for measuring noise factor.*

Next, the generator V_s is turned on, and an input signal is applied until the output power doubles (output rms voltage increases by 3 dB over that previously measured). Under these conditions the following equation is satisfied

$$(AV_s)^2 + (V_{no})^2 = 2V_{no}^2, \tag{9-11}$$

therefore,

$$AV_s = V_{no}. \tag{9-12}$$

Substituting Eq. 9-12 into Eq. 9-7 gives

$$F = \left(\frac{V_s}{V_t}\right)^2 . \tag{9-13}$$

Substituting from Eq. 9-5 for V_t produces

$$F = \frac{V_s^2}{1.6 \times 10^{-20} BR_s} . \tag{9-14}$$

Since the noise factor is not a function of R_L, any value of load resistor can be used for the measurement.

The disadvantage of this method is that the noise bandwidth of the device* must be known.

A better method of measuring noise factor is to use a noise diode as a white noise source. The measuring circuit is shown in Fig. 9-3. I_{dc} is the direct current through the noise diode, and R_s is the source resistance. The shot noise in the diode is

$$I_{sh} = \sqrt{3.2 \times 10^{-19} I_{dc} B} \ . \tag{9-15}$$

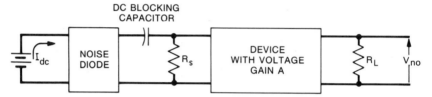

Figure 9-3. *Noise diode method of measuring noise factor.*

Using Thevenin's theorem, the shot-noise current generator can be replaced by a voltage generator V_{sh} in series with R_s, where

$$V_{sh} = I_{sh} R_s. \tag{9-16}$$

The rms noise voltage output V_{no} is first measured with the diode current equal to zero. This voltage consists of two parts: that due to the thermal noise of the source resistor, and that due to the noise in the device. Therefore,

$$V_{no} = \sqrt{(AV_t)^2 + (\text{Device Noise})^2} \ . \tag{9-17}$$

Diode current is then allowed to flow and is increased until the output noise power doubles (output rms voltage increases by 3dB). Under these conditions the following equation is satisfied:

$$(AV_{sh})^2 + (V_{no})^2 = 2(V_{no})^2, \tag{9-18}$$

therefore,

$$V_{no} = AV_{sh} = A\,I_{sh}\,R_s. \tag{9-19}$$

*It should be remembered that the noise bandwidth is usually not equal to the 3-dB bandwidth (see Chapter 8).

Substituting V_{no} from Eq. 9-19 into Eq. 9-7, gives

$$F = \frac{(I_{sh} R_s)^2}{V_t^2} .$$
(9-20)

Substituting Eqs. 9-15 and 9-5 for I_{sh} and V_t respectively, gives

$$F = 20 I_{dc} R_s.$$
(9-21)

The noise factor is now a function of only the direct current through the diode, and the value of the source resistance. Both of these quantities are easily measured. Neither the gain nor the noise bandwidth of the device need be known.

CALCULATING S/N RATIO AND INPUT NOISE VOLTAGE FROM NOISE FACTOR

Once noise factor is known, it can be used to calculate the signal-to-noise ratio and the input noise voltage. For these calculations it is important that the source resistance used in the circuit be the same as that used to make the noise factor measurement. Rearranging Eq. 9-8 gives

$$V_{no} = A \sqrt{4kTBR_s F} .$$
(9-22)

If the input signal is V_s, the output signal voltage is $V_o = A V_s$. Therefore, the output signal-to-noise power ratio is

$$\frac{S_o}{N_o} = \frac{P_{(signal)}}{P_{(noise)}} ,$$
(9-23)

or

$$\frac{S_o}{N_o} = \left(\frac{A V_s}{V_{no}} \right)^2 .$$
(9-24)

Using Eq. 9-22 to substitute for V_{no},

$$\frac{S_o}{N_o} = \frac{(V_s)^2}{4kTBR_s F} .$$
(9-25)

Signal-to-noise ratio, as used in Eqs. 9-23, 9-24, and 9-25, refers to a power ratio. However, signal-to-noise is sometimes expressed as a voltage ratio. Care should be taken as to whether a specified signal-to-noise ratio is a power or voltage ratio, since the two are not numerically equal. When expressed in decibels, the power signal-to-noise ratio is $10 \log (S_o/N_o)$.

Another useful quantity is the total equivalent input noise voltage (V_{nt}), which is the output noise voltage (Eq. 9-22) divided by the gain

$$V_{nt} = \frac{V_{no}}{A} = \sqrt{4kTBR_s F} \ . \tag{9-26}$$

The total equivalent input noise voltage is a single noise source that represents the total noise in the circuit. *For optimum noise performance, V_{nt} should be minimized.* Minimizing V_{nt} is equivalent to maximizing the signal-to-noise ratio provided the signal voltage remains constant. This is discussed further in the section on optimum source resistance.

The equivalent input noise voltage consists of two parts, one due to the thermal noise of the source and the other due to the device noise.

Representing the device noise by V_{nd}, we can write the total equivalent input noise voltage as

$$V_{nt} = \sqrt{(V_t)^2 + (V_{nd})^2} \ , \tag{9-27}$$

where V_t is the open circuit thermal noise voltage of the source resistance. Solving Eq. 9-27 for V_{nd} gives

$$V_{nd} = \sqrt{(V_{nt})^2 - (V_t)^2} \ . \tag{9-28}$$

Substituting Eqs. 9-4 and 9-26 into Eq. 9-28 gives

$$V_{nd} = \sqrt{4kTBR_s (F-1)} \ . \tag{9-29}$$

NOISE VOLTAGE AND CURRENT MODEL

A more recent approach, and one that overcomes the limitations of noise factor, is to model the noise in terms of an equivalent noise voltage and current. The actual network can be modeled as a noise-free device with two noise generators, V_n and I_n, connected to the input side of a network, as shown in Fig. 9-4. V_n represents the device noise that exists when R_s equals zero, and I_n represents the additional device noise that occurs when R_s does not equal zero. The use of these two noise generators plus a complex correlation coefficient (not shown) completely characterizes the noise performance of the device (Rothe and Dahlke, 1956). Although V_n

and I_n are normally correlated to some degree, values for the correlation coefficient are seldom given on manufacturers data sheets. In addition the typical spread of values of V_n and I_n for a device normally overshadows the effect of the correlation coefficient. Therefore, it is common practice to assume the correlation coefficient is equal to zero. This will be done in the remainder of this chapter.

Figure 9-5 shows representative curves of noise voltage and noise current. As can be seen in Fig. 9-5, the data normally consists of a plot of $V_n/\sqrt{B}$ and $I_n/\sqrt{B}$ versus frequency. The noise voltage or current over a band of frequencies can be found by integrating $[V_n/\sqrt{B}\,]^2$ or $[I_n/\sqrt{B}\,]^2$ versus frequency and then taking the square root of the result. In the case when $V_n/\sqrt{B}$ or $I_n/\sqrt{B}$ is constant over the desired bandwidth the total noise voltage or current can be found simply by multiplying $V_n/\sqrt{B}$ or $I_n/\sqrt{B}$ by the square root of the bandwidth.

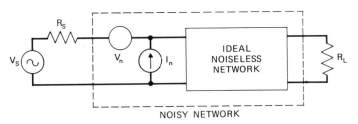

Figure 9-4. *A noisy network modeled by the addition of an input noise voltage and current source.*

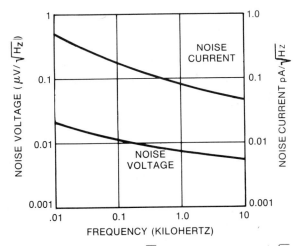

Figure 9-5. *Typical noise voltage $V_n/\sqrt{B}$ and noise current $I_n/\sqrt{B}$ curves.*

Using these curves and the equivalent circuit of Fig. 9-4, the total equivalent input noise voltage, signal-to-noise ratio, or noise factor for any circuit can be determined. This can be done for any source impedance, resistive or reactive, and across any frequency spectrum. The device must, however, be operated at or near the bias conditions for which the curves are specified. Quite often, additional curves are given showing the variation of these noise generators with bias points. With a set of these curves, the noise performance of the device is completely specified under all operating conditions.

The representation of noise data in terms of the equivalent parameters V_n and I_n can be used for any device. Field effect transistors and op-amps are usually specified in this manner. Some bipolar transistor manufacturers are also beginning to use the V_n–I_n parameters instead of noise factor.

The total equivalent input noise voltage of a device is an important parameter. Assuming no correlation between noise sources, this voltage, which combines the affect of V_n, I_n, and the thermal noise of the source can be written as

$$V_{nt} = \sqrt{4kTBR_s + V_n^2 + (I_n R_s)^2} \, , \tag{9-30}$$

where V_n and I_n are the noise voltage and noise current over the bandwidth B. For optimum noise performance the total noise voltage represented by Eq. 9-30 should be minimized. This is discussed further in the optimum source resistance section.

The total equivalent input voltage per square root of bandwidth can be written as

$$\frac{V_{nt}}{\sqrt{B}} = \sqrt{4kTR_s + \left(\frac{V_n}{\sqrt{B}}\right)^2 + \left(\frac{I_n R_s}{\sqrt{B}}\right)^2} \, . \tag{9-31}$$

The equivalent input noise voltage due to device noise only can be calculated by subtracting the thermal noise component from Eq. 9-30. The equivalent input device noise then becomes

$$V_{nd} = \sqrt{V_n^2 + (I_n R_s)^2} \, . \tag{9-32}$$

Figure 9-6 is a plot of the total equivalent noise voltage per square root of the bandwidth for a typical low-noise bipolar transistor, junction field-effect transistor, and op-amp. The thermal noise voltage generated by the source resistance is also shown. The thermal noise curve places a lower limit on the total input noise voltage. As can be seen from this figure, when the source resistance is between 10,000 ohms and 1 megohm this FET has a total noise voltage only slightly greater than the thermal noise in the source resistance. On the basis of noise, this FET approaches an ideal device when the source resistance is in this range. With low source resistance, however, a bipolar transistor generally has less noise than an FET. In most cases, the op-amp has more noise than either of the other devices. The reasons for this are discussed in the section on op-amp noise.

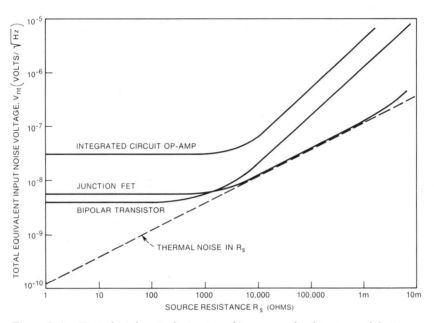

Figure 9-6. *Typical total equivalent noise voltage curves for three types of devices.*

MEASUREMENT OF V_n AND I_n

It is a relatively simple matter to measure the parameters V_n and I_n for a device. The method can best be described by referring to Fig. 9-4 and recalling from Eq. 9-30 that the total equivalent noise voltage V_{nt} is

$$V_{nt} = \sqrt{4kTBR_s + V_n^2 + (I_n R_s)^2} \ . \tag{9-33}$$

To determine V_n, the source resistance is set equal to zero, causing two terms in Eq. 9-33 to equal zero, and the output noise voltage V_{no} is measured. If the voltage gain of the circuit is A,

$$V_{no} = A V_{nt} = A V_n, \qquad \text{for } R_s = 0. \tag{9-34}$$

The equivalent input noise voltage is

$$V_n = \frac{V_{no}}{A}, \qquad \text{for } R_s = 0. \tag{9-35}$$

To measure I_n, a second measurement is made with a very large source resistance. The source resistance should be large enough so that the first two terms in Eq. 9-33 are negligible. This will be true if the measured output noise voltage V_{no} is

$$V_{no} \gg A\sqrt{4kTBR_s + V_n^2}\ .$$

Under these conditions the equivalent input noise current is

$$I_n = \frac{V_{no}}{AR_s}, \qquad \text{for } R_s \text{ large.} \tag{9-36}$$

CALCULATING NOISE FACTOR AND S/N RATIO FROM V_n-I_n

Knowing the equivalent input noise voltage V_n, the current I_n, and the source resistance R_s, the noise factor can be calculated by referring to Fig. 9-4. This derivation is left as a problem in Appendix D. The result is

$$F = 1 + \frac{1}{4kTB}\left(\frac{V_n^2}{R_s} + I_n^2 R_s\right), \tag{9-37}$$

where V_n and I_n are the equivalent input noise voltage and current over the bandwidth B of interest.

The value of R_s producing the minimum noise factor can be determined from Eq. 9-37 by differentiating it with respect to R_s. The resulting R_s for minimum noise factor is

$$R_{so} = \frac{V_n}{I_n}. \tag{9-38}$$

If Eq. 9-38 is substituted back into Eq. 9-37 the minimum noise factor can be determined, and is

$$F_{\min} = 1 + \frac{V_n I_n}{2kTB}. \tag{9-39}$$

The output power signal-to-noise ratio can also be calculated from the circuit of Fig. 9-4. This derivation is left as a problem in Appendix D. The result is

$$\frac{S_o}{N_o} = \frac{(V_s)^2}{(V_n)^2 + (I_n R_s)^2 + 4kTBR_s}, \tag{9-40}$$

where V_s is the input signal voltage.

For constant V_s maximum signal-to-noise ratio occurs when $R_s = 0$, and is

$$\left.\frac{S_o}{N_o}\right|_{max} = \left(\frac{V_s}{V_n}\right)^2. \tag{9-41}$$

It should be noted that when V_s is constant and R_s is variable minimum noise factor occurs when $R_s = V_n/I_n$, but maximum signal-to-noise ratio occurs at $R_s = 0$. Minimum noise factor, therefore, does not necessarily represent maximum signal-to-noise ratio or minimum noise. This can best be understood by referring to Fig. 9-7, which is a plot of the total equivalent input noise voltage V_{nt} for a typical device. When $R_s = V_n/I_n$, the ratio of the device noise to the thermal noise is a minimum. However,

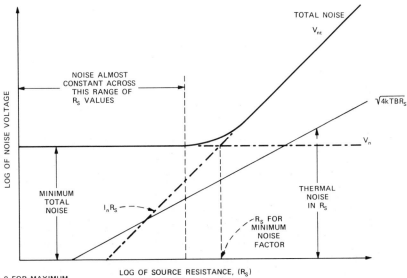

Figure 9-7. *Total equivalent input noise voltage V_{nt} for a typical device. The total noise voltage is made up of three components (thermal noise, V_n, and $I_n R_s$) as was shown in Eq. 9-30.*

the device noise and the thermal noise are both minimum when $R_s = 0$. Although minimum equivalent input noise voltage (and maximum signal-to-noise ratio) occurs mathematically at $R_s = 0$, there is actually a range of values of R_s over which it is almost constant, as shown in Fig. 9-7. In this range, V_n of the device is the predominant noise source. For large values of source resistance, I_n is the predominant noise source.

OPTIMUM SOURCE RESISTANCE

Since the maximum signal-to-noise ratio occurs at $R_s = 0$ and minimum noise factor occurs at $R_s = V_n/I_n$ the question of what is the optimum source resistance for the best noise performance arises. The requirement of a zero resistance source is impractical since all actual sources have a finite source resistance. However, as was shown in Fig. 9-7, as long as R_s is small there is a range of values over which the total noise voltage is almost constant.

In practice the circuit designer does not always have control over the source resistance. A source of fixed resistance is used for one reason or another. The question then arises as to whether this source resistance should be transformed to the value that produces minimum noise factor. The answer to this question depends on how the transformation is made.

If the actual source resistance is less than $R_s = V_n/I_n$, a physical resistor should not be inserted in series with R_s to increase the resistance. To do this would produce three detrimental effects:

1. It increases the thermal noise due to the larger source resistance. (This increase is proportional to $\sqrt{R}$.)

2. It increases the noise due to the current from the input noise current generator flowing through the larger resistor. (This increase is proportional to R.)

3. It decreases the amount of the signal getting to the amplifier.

The noise performance can, however, be improved by using a transformer to effectively raise the value of R_s to a value closer to $R_s = V_n/I_n$, thus minimizing the noise produced by the device. At the same time, the signal voltage is stepped up by the turns ratio of the transformer. This effect is cancelled by the fact that the thermal noise voltage of the source resistance is also stepped up by the same factor. There is, however, a net increase in signal-to-noise ratio when this is done.

If the actual source resistance is greater than that required for minimum noise factor, noise performance can still be improved by transforming the higher value of R_s to a value closer to $R_s = V_n/I_n$. The noise will, however, be greater than if a lower impedance source was used.

For optimum noise performance, the lowest possible source impedance should be used. Once this is decided, noise performance can be further improved by transformer coupling this source to match the impedance $R_s = V_n / I_n$.

The improvement in signal-to-noise ratio that is possible by using a transformer can best be seen by rewriting Eq. 9-3 as

$$\frac{S_o}{N_o} = \frac{1}{F}\left(\frac{S_i}{N_i}\right).$$

(9-42)

Assuming a fixed source resistance, adding an ideal transformer of any turns ratio does not change the input signal-to-noise ratio. With the input signal-to-noise ratio fixed, the output signal-to-noise ratio will be maximized when the noise factor F is a minimum. F is a minimum when the device sees a source resistance $R_s = V_n / I_n$. Therefore, transformer coupling the actual source resistance minimizes F and maximizes the output signal-to-noise ratio. If the value of the source resistance is not fixed, choosing R_s to minimize F does not necessarily produce optimum noise performance. However, for a given source resistance R_s, the least noisy circuit is the one with the smallest F.

When using transformer coupling, thermal noise in the transformer winding must be accounted for. This can be done by adding to the source resistance the primary winding resistance, plus the secondary winding resistance divided by the square of the turns ratio. The turns ratio is defined as the number of turns of the secondary divided by the number of turns of the primary. Despite this additional noise introduced by the transformer, the signal-to-noise ratio is normally increased sufficiently to justify using the transformer if the actual source resistance is more than an order of magnitude different than the optimum source resistance.

Another source of noise to consider when using a transformer is its sensitivity to pickup from magnetic fields. Shielding the transformer is often necessary to reduce this pickup to an acceptable level.

The improvement in signal-to-noise ratio due to transformer coupling can be expressed in terms of the signal-to-noise improvement factor (SNI) defined as

$$SNI = \frac{(S/N) \text{ using transformer}}{(S/N) \text{ without transformer}}.$$

(9-43)

It can be shown that the signal-to-noise improvement factor can also be expressed in a more useful form as

$$SNI = \frac{(F) \text{ without transformer}}{(F) \text{ with transformer}}. \tag{9-44}$$

NOISE FACTOR OF CASCADED STAGES

Signal-to-noise ratio and total equivalent input noise voltage should be used in designing the components of a system for optimum noise performance. Once the components of a system have been designed, it is usually advantageous to express the noise performance of the individual components in terms of noise factor. The noise factor of the various components can then be combined as follows.

The overall noise factor of a series of networks connected in cascade (see Fig. 9-8) was shown by Friis (1944) to be

$$F = F_1 + \frac{F_2 - 1}{G_1} + \frac{F_3 - 1}{G_1 G_2} + \cdots + \frac{F_m - 1}{G_1 G_2 \cdots G_{m-1}}. \tag{9-45}$$

where F_1 and G_1 are the noise factor and available power gain* of the first stage, F_2, G_2 are those of the second stage.

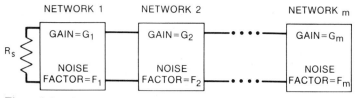

Figure 9-8. *Networks in cascade.*

Equation 9-45 clearly shows the important fact that: *with sufficient gain G_1 in the first stage of a system, the total noise factor is primarily determined by the noise factor F_1 of the first stage.*

Example 9-1. Figure 9-9 shows a number of identical amplifiers operating in cascade on a transmission line. Each amplifier has an available power gain G, and the amplifiers are spaced so the loss in the section of cable between amplifiers is also G. This type of arrangement can be used in a telephone trunk circuit or a CATV distribution system. The amplifiers all

---

*$G = A^2 R_s / R_o$ where A is the open-circuit voltage gain (open-circuit output voltage divided by source voltage). R_s is the source resistance, and R_o is the network output impedance.

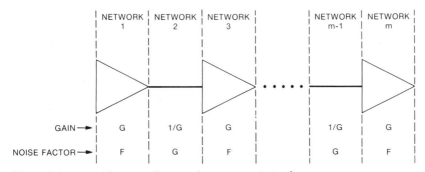

Figure 9-9. *Amplifiers equally spaced on a transmission line.*

have an available power gain equal to G and a noise factor F. The cable sections have an insertion gain $1/G$ and a noise factor G.* Equation 9-45 then becomes

$$F_t = F + \frac{G-1}{G} + \frac{F-1}{1} + \frac{G-1}{G} + \frac{F-1}{1} + \cdots + \frac{F-1}{1}, \quad (9\text{-}46)$$

$$F_t = F + 1 - \frac{1}{G} + F - 1 + 1 - \frac{1}{G} + F - 1 + \cdots + F - 1. \quad (9\text{-}47)$$

For K amplifiers and $K-1$ cable sections,

$$F_t = KF - \frac{K-1}{G}. \quad (9\text{-}48)$$

If $FG \gg 1$,

$$F_t = KF. \quad (9\text{-}49)$$

The overall noise figure equals

$$(NF)_t = 10 \log F + 10 \log K. \quad (9\text{-}50)$$

The overall noise figure, therefore, equals the noise figure of the first amplifier plus ten times the logarithm of the number of stages. Another way of looking at this is that every time the number of stages is doubled, the noise figure increases by 3 dB. This limits the maximum number of amplifiers that can be cascaded.

*This can be derived by applying the basic noise factor definition (Eq. 9-1) to the cable section. The cable is considered a matched transmission line operating at its characteristic impedance.

Example 9-2. Figure 9-10 shows an antenna connected to a TV set by a section of 300 Ω matched transmission line. If the transmission line has 6 dB of insertion loss and the TV set has a noise figure of 14 dB, what signal voltage is required at the antenna terminal for a 40-dB signal-to-noise ratio at the terminals of the TV set? To solve this problem, all the noise sources in the system are converted to equivalent noise voltages at one point, in this case the input to the TV set. The noise voltages can then be combined, and the appropriate signal level needed to produce the required signal-to-noise ratio can be calculated.

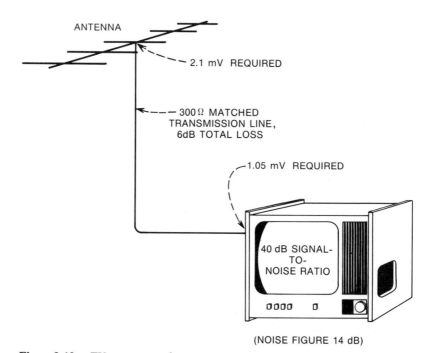

ANTENNA

2.1 mV REQUIRED

300 Ω MATCHED TRANSMISSION LINE, 6dB TOTAL LOSS

1.05 mV REQUIRED

40 dB SIGNAL- TO- NOISE RATIO

(NOISE FIGURE 14 dB)

Figure 9-10. *TV set connected to antenna.*

The thermal noise at the input of the TV set due to a 300-Ω input impedance with a 4-MHz bandwidth is −53.5 dBmV (2.1 μV).* Since the TV set adds 14 dB of noise to the input thermal noise, the total input noise level is −39.5 dBmV (thermal noise voltage in dB + noise figure). Since a signal-to-noise ratio of 40 dB is required, the signal voltage at the amplifier

---

*The open circuit noise voltage for a 300-Ω resistor and a 4-MHz bandwidth is 4.2 μV. When this source is connected to a 300-Ω load, it delivers one-half this voltage, or 2.1 μV, to the load.

input must be $+0.5$ dBmV (total input noise in dB + signal-to-noise ratio in dB). The transmission line has 6 dB of loss, so therefore, the signal voltage at the antenna terminal must be $+6.5$ dBmV or 2.1 mV. To be able to add terms directly, as in this example, all the quantities must be referenced to the same impedance level, in this case 300 Ω.

NOISE TEMPERATURE

Another method of specifying noise performance of a circuit or device is by the concept of equivalent input noise temperature (T_e).

The equivalent input noise temperature of a circuit can be defined as the increase in source resistance temperature necessary to produce the observed noise power at the output of the circuit. The standard reference temperature T_0 for noise temperature measurements is 290°K.

Figure 9-11 shows a noisy amplifier with a source resistance R_s at temperature T_0. The total measured output noise is V_{no}. Figure 9-12 shows an ideal noiseless amplifier having the same gain as the amplifier in Fig. 9-11, and also showing a source resistance R_s. The temperature of the

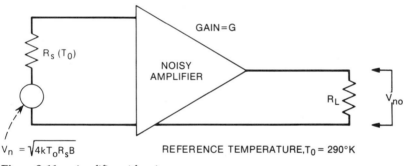

$$V_n = \sqrt{4kT_0 R_s B}$$

REFERENCE TEMPERATURE, T_0 = 290°K

Figure 9-11. *Amplifier with noise.*

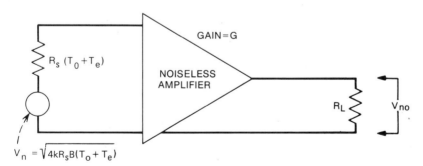

$$V_n = \sqrt{4kR_s B(T_0 + T_e)}$$

Figure 9-12. *Source resistance temperature increased to account for amplifier noise.*

source resistance is now increased by T_e so the total measured output noise V_{no} is the same as in Fig. 9-11. T_e is then the equivalent noise temperature of the amplifier.

The equivalent input noise temperature is related to the noise factor F by

$$T_e = 290(F-1),$$ (9-51)

and to noise figure NF by

$$T_e = 290(10^{NF/10} - 1).$$ (9-52)

In terms of the equivalent input noise voltage and current ($V_n - I_n$), the noise temperature can be written as

$$T_e = \frac{V_n^2 + (I_n R_s)^2}{4kBR_s}.$$ (9-53)

The equivalent input noise temperature of a number of amplifiers in cascade can be shown to be

$$T_{e(total)} = T_{e1} + \frac{T_{e2}}{G_1} + \frac{T_{e3}}{G_1 G_2} + \cdots,$$ (9-54)

where T_{e1} and G_1 are the equivalent input noise temperature and available power gain of the first stage, T_{e2} and G_2 are the same for the second stage, and so on.

BIPOLAR TRANSISTOR NOISE

The noise figure versus frequency for a typical bipolar transistor is shown in Fig. 9-13. It can be seen that the noise figure is constant across some middle range of frequencies and rises on both sides. The low frequency increase in noise figure is due to "$1/f$" or contact noise (see Chapter 8). The $1/f$ noise and the frequency f_1 increase with increasing collector current.

Above frequency f_1, the noise is due to white noise sources consisting of thermal noise in the base resistance and shot noise in the emitter and collector junctions. The white noise sources can be minimized by choosing a transistor with small base resistance, large current gain, and high alpha cutoff frequency. The increase in noise figure at frequencies above f_2 is due to: (1) the decrease in transistor gain at these frequencies, and (2) the transistor noise produced in the output (collector) junction which, therefore, is not affected by transistor gain.

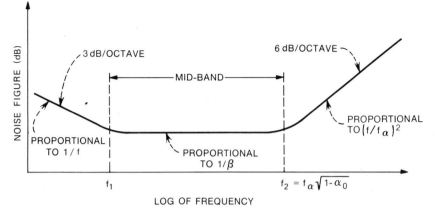

Figure 9-13. *Noise figure versus frequency for bipolar transistor.*

For a typical audio transistor the frequency f_1, below which the noise begins to increase, may be between 1 and 50 kHz. The frequency f_2, above which the noise increases, is usually greater than 10 MHz. In transistors designed for RF use, f_2 may be much higher.

Transistor Noise Factor

The theoretical expression for bipolar transistor noise factor can be derived by starting with the T-equivalent circuit of a transistor, as shown in Fig. 9-14, neglecting the leakage term I_{CBO}. By neglecting $r_c (r_c \gg R_L)$ and adding the following noise sources, (1) thermal noise of the base resistance, (2) shot noise in emitter diode, (3) shot noise in collector, and (4) thermal noise in the source resistance, the circuit can be revised to form the equivalent circuit shown in Fig. 9-15.

The noise factor can be obtained from the circuit in Fig. 9-15 and the relationships;

$$I_c = \alpha_o I_e, \tag{9-55}$$

$$r_e = \frac{kT}{qI_e} \approx \frac{26}{I_e \text{(ma)}}, \tag{9-56}$$

$$|\alpha| = \frac{|\alpha_o|}{\sqrt{1 + \left(\dfrac{f}{f_\alpha}\right)^2}}, \tag{9-57}$$

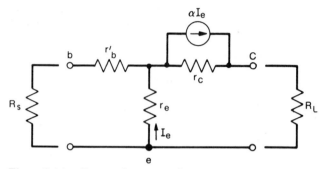

Figure 9-14. *T-equivalent circuit for a transistor.*

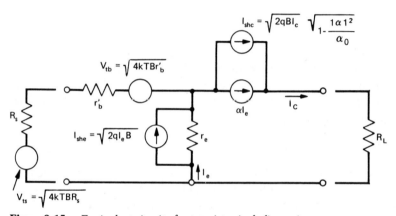

Figure 9-15. *Equivalent circuit of a transistor including noise sources.*

where α_o is the dc value of the transistor common-base current gain alpha, k is Boltzmann's constant, q is the electron charge, f_α is the transistor alpha cutoff frequency, and f is the frequency variable. Using this equivalent circuit, Nielsen (1957) has shown the noise factor of the transistor to be

$$F = 1 + \frac{r_b'}{R_s} + \frac{r_e}{2R_s} + \frac{(r_e + r_b' + R_s)^2}{2r_e R_s \beta_o}\left[1 + \left(\frac{f}{f_\alpha}\right)^2 (1 + \beta_o)\right], \quad (9\text{-}58)$$

where β_o is the dc value of the common-emitter current gain beta,

$$\beta_o = \frac{\alpha_o}{1 - \alpha_o}. \quad (9\text{-}59)$$

This equation does not include the effect of the $1/f$ noise, and is valid at all frequencies above f_1 in Fig. 9-13. The $1/f$ noise can be represented as an additional noise current source in parallel with αI_e in the collector circuit.

The second term in Eq. 9-58 represents the thermal noise in the base, the third term represents shot noise in the emitter, and the fourth term represents shot noise in the collector. This equation is applicable to both the common emitter and the common base configurations.

The value of source resistance R_{so} for minimum noise factor can be determined by differentiating Eq. 9-58 with respect to R_s and setting the result equal to zero. This source resistance is found to be

$$R_{so} = \left[(r_b' + r_e)^2 + \frac{(2r_b' + r_e)\beta_o r_e}{1 + (f/f_\alpha)^2 (1 + \beta_o)} \right]^{1/2}. \tag{9-60}$$

For most bipolar transistors, the value of source resistance for minimum noise factor is close to the value that produces maximum power gain. Most transistor applications operate the transistor at a frequency considerably below the alpha cutoff frequency. Under this condition ($f \ll f_\alpha$), assuming $\beta_o \gg 1$, Eq. 9-60 reduces to

$$R_{so} = \sqrt{(2r_b' + r_e)\beta_o r_e}. \tag{9-61}$$

If, in addition, the base resistance r_b' is negligible (not always the case), Eq. 9-61 becomes

$$R_{so} \approx r_e \sqrt{\beta_o}. \tag{9-62}$$

This equation is also useful for making quick approximations of the source resistance that produces minimum noise factor. Equation 9-62 shows that the higher the common emitter current gain β_o of the transistor, the higher the value of R_{so}.

$V_n - I_n$ For Transistor

To determine the parameters for the equivalent input noise voltage and current model, we must first determine the total equivalent input noise voltage V_{nt}. Substituting Eq. 9-58 into Eq. 9-26 and squaring the result, gives

$$V_{nt}^2 = 2kTB(r_e + 2r_b' + 2R_s) + \frac{2kTB(r_e + r_b' + R_s)^2}{r_e \beta_o} \left[1 + \left(\frac{f}{f_\alpha} \right)^2 (1 + \beta_o) \right]. \tag{9-63}$$

The equivalent input noise voltage squared V_n^2 is obtained by making $R_s = 0$ in Eq. 9-63 (see Eqs. 9-34 and 9-35), giving

$$V_n^2 = 2kTB\,(r_e + 2r_b') + \frac{2kTB\,(r_e + r_b')^2}{r_e\beta_o}\left[1 + \left(\frac{f}{f_\alpha}\right)^2(1 + \beta_o)\right]. \quad (9\text{-}64)$$

To determine I_n^2 we must divide Eq. 9-63 by R_s^2 and then make R_s large (see Eqs. 9-34 and 9-36), giving

$$I_n^2 = \frac{2kTB}{r_e\beta_o}\left[1 + \left(\frac{f}{f_\alpha}\right)^2(1 + \beta_o)\right]. \quad (9\text{-}65)$$

JUNCTION FIELD EFFECT TRANSISTOR NOISE

There are three important noise mechanisms in a junction FET: (1) the shot noise produced in the reverse biased gate, (2) the thermal noise generated in the channel between source and drain, and (3) the $1/f$ noise generated in the space charge region between gate and channel.

Figure 9-16 is the noise equivalent circuit for a junction FET. Noise generator I_{sh} represents the shot noise in the gate circuit and generator I_{tc} represents the thermal noise in the channel. I_{ts} is the thermal noise of the source admittance G_s. The FET has an input admittance g_{11}, and a forward transconductance g_{fs}.

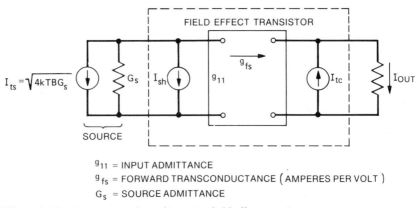

g_{11} = INPUT ADMITTANCE
g_{fs} = FORWARD TRANSCONDUCTANCE (AMPERES PER VOLT)
G_s = SOURCE ADMITTANCE

Figure 9-16. *Noise equivalent of junction field effect transistor.*

FET Noise Factor

Assuming no correlation between I_{sh} and $I_{tc}*$ in Fig. 9-16, the total output noise current can be written as

$$I_{out} = \left[\frac{4kTBG_s g_{fs}^2}{(G_s + g_{11})^2} + \frac{I_{sh}^2 g_{fs}^2}{(G_s + g_{11})^2} + I_{tc}^2 \right]^{1/2}. \tag{9-66}$$

The output noise current due to the thermal noise of the source only is

$$I_{out}(source) = \left[\frac{\sqrt{4kTBG_s}}{G_s + g_{11}} \right] g_{fs}. \tag{9-67}$$

The noise factor F is Eq. 9-66 squared, divided by Eq. 9-67 squared, or

$$F = 1 + \frac{I_{sh}^2}{4kTBG_s} + \frac{I_{tc}^2}{4kTBG_s(g_{fs})^2}(G_s + g_{11})^2. \tag{9-68}$$

I_{sh} is the input shot noise and equals

$$I_{sh} = \sqrt{2qI_{gss}B}, \tag{9-69}$$

where I_{gss} is the total gate leakage current. I_{tc} is the thermal noise of the channel and equals

$$I_{tc} = \sqrt{4kTBg_{fs}}. \tag{9-70}$$

Substituting Eqs. 9-69 and 9-70 into 9-68 and recognizing that

$$\frac{2q}{4kT}I_{gss} = g_{11} \tag{9-71}$$

gives for the noise factor

$$F = 1 + \frac{g_{11}}{G_s} + \frac{1}{G_s g_{fs}}(G_s + g_{11})^2. \tag{9-72}$$

Rewriting Eq. 9-72 in terms of the resistances instead of admittances gives

$$F = 1 + \frac{R_s}{r_{11}} + \frac{R_s}{g_{fs}}\left(\frac{1}{R_s} + \frac{1}{r_{11}} \right)^2. \tag{9-73}$$

*At high frequencies, noise generators I_{sh} and I_{tc} show some correlation. As a practical matter, however, this is normally neglected.

Neither Eq. 9-72 nor Eq. 9-73 include the effect of the $1/f$ noise. The second term in the equations represent the contribution from the shot noise in the gate junction. The third term represents the contribution of the thermal noise in the channel.

For low noise operation, an FET should have high gain (large g_{fs}) and a high input resistance r_{11} (small gate leakage).

Normally, at low frequencies, the source resistance R_s is less than the gate leakage resistance r_{11}. Under these conditions Eq. 9-73 becomes

$$F \approx 1 + \frac{1}{g_{fs} R_s}. \qquad (9\text{-}74)$$

In the case of an insulated gate FET (IGFET) or metal oxide FET (MOSFET) there is no p–n gate junction and therefore no shot noise, so Eq. 9-74 applies. However, in the cases of IGFETs or MOSFETs the $1/f$ noise is often greater than in the case of JFETs.

$V_n - I_n$ Representation of FET

The total equivalent input noise voltage can be obtained by substituting Eq. 9-73 into Eq. 9-26, giving

$$V_{nt}^2 = 4kTBR_s \left[1 + \frac{R_s}{r_{11}} + \frac{R_s}{g_{fs}} \left(\frac{1}{R_s} + \frac{1}{r_{11}} \right)^2 \right]. \qquad (9\text{-}75)$$

Making $R_s = 0$ in Eq. 9-75 gives the equivalent input noise voltage squared (see Eqs. 9-34 and 9-35) as

$$V_n^2 = \frac{4kTB}{g_{fs}}. \qquad (9\text{-}76)$$

To determine I_n^2 we must divide Eq. 9-75 by R_s^2 and then make R_s large (see Eqs. 9-34 and 9-36), giving

$$I_n^2 = \frac{4kTB\left(1 + g_{fs} r_{11}\right)}{g_{fs} r_{11}^2}. \qquad (9\text{-}77)$$

For the case when $g_{fs} r_{11} \gg 1$, Eq. 9-77 becomes

$$I_n^2 = \frac{4kTB}{r_{11}}. \qquad (9\text{-}78)$$

NOISE IN IC OPERATIONAL AMPLIFIERS

The input stage of an operational amplifier is of primary concern in determining the noise performance of the device. Most monolithic op-amps use a differential input configuration that uses two and sometimes four input transistors. Figure 9-17 shows a simplified schematic of a typical two-transistor input circuit used in an operational amplifier. Since two input transistors are used, the noise voltage is approximately $\sqrt{2}$ times that for a single-transistor input stage. In addition, some monolithic transistors have lower current gains (β) than discrete transistors, and that also increases the noise.

Therefore, in general, operational amplifiers are inherently higher noise devices than discrete transistor amplifiers. This can be seen in the typical equivalent input noise voltage curves shown in Fig. 9-6. A discrete bipolar transistor stage preceding an op-amp can often provide lower noise performance along with the other advantages of the operational amplifier. Op-amps do have the advantage of a balanced input with low temperature drift and low-input offset currents.

The noise characteristics of an operational amplifier can best be modeled by using the equivalent input noise voltage and current $V_n - I_n$. Figure 9-18A shows a typical operational amplifier circuit. Figure 9-18B shows this same circuit with the equivalent noise voltage and current sources included.

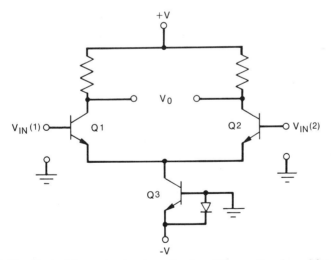

Figure 9-17. *Typical input circuit schematic of an IC operational amplifier. Transistor Q_3 acts as a constant current source to provide dc bias for the input transistors Q_1 and Q_2.*

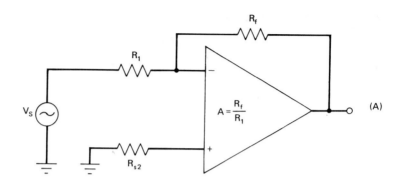

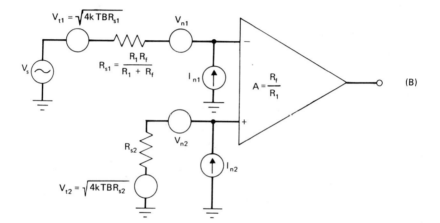

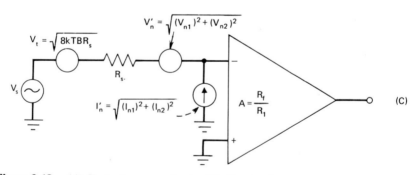

Figure 9-18. *(A) Typical op-amp circuit; (B) Circuit of A with noise sources added; (C) Circuit of B with noise sources combined at one terminal for the case $R_{s1} = R_{s2} = R_s$.*

240

The equivalent circuit in Fig. 9-18B can be used to calculate the total equivalent input noise voltage, which is

$$V_{nt} = \left[4kTB\left(R_{s1} + R_{s2}\right) + V_{n1}^2 + V_{n2}^2 + \left(I_{n1}R_{s1}\right)^2 + \left(I_{n2}R_{s2}\right)^2 \right]^{1/2}. \quad (9\text{-}79)$$

It should be noted that V_{n1}, V_{n2}, I_{n1}, and I_{n2} are also functions of the bandwidth B.

The two noise voltage sources of Eq. 9-66 can be combined by defining

$$(V_n')^2 = V_{n1}^2 + V_{n2}^2. \qquad (9\text{-}80)$$

Equation 9-79 can then be rewritten as

$$V_{nt} = \left[4kTB\left(R_{s1} + R_{s2}\right) + \left(V_n'\right)^2 + \left(I_{n1}R_{s1}\right)^2 + \left(I_{n2}R_{s2}\right)^2 \right]^{1/2}. \quad (9\text{-}81)$$

Although the voltage sources have been combined, the two noise current sources are still required in Eq. 9-81. If, however, $R_{s1} = R_{s2}$, which is usually the case since this minimizes the dc output offset voltage due to input bias current, then the two noise current generators can be combined by defining

$$(I_n')^2 = I_{n1}^2 + I_{n2}^2. \qquad (9\text{-}82)$$

For $R_{s1} = R_{s2} = R_s$, Eq. 9-81 reduces to

$$V_{nt} = \left[8kTBR_s + \left(V_n'\right)^2 + \left(I_n'R_s\right)^2 \right]^{1/2}. \qquad (9\text{-}83)$$

The equivalent circuit for this case is shown in Fig. 9-18C. To obtain optimum noise performance (maximum signal-to-noise ratio) from an op-amp, the total equivalent input noise voltage V_{nt} should be minimized.

Methods of Specifying Op-Amp Noise

Various methods are used by op-amp manufacturers to specify noise for their devices. Sometimes they provide values for V_n and I_n at each input terminal, as represented by the equivalent circuit in Fig. 9-19A. Due to the symmetry of the input circuit, the noise voltage and noise current at each input are equal. A second method is to provide combined values, V_n' and I_n', which are then applied to one input only, as shown in Fig. 9-19B. To combine the two noise current generators, it must be assumed that equal source resistors are connected to the two input terminals. The magnitudes of the combined noise voltage generators in Fig. 9-19B, with respect to the individual generators in Fig. 9-19A, are

$$V_n' = \sqrt{2}\, V_n, \qquad (9\text{-}84)$$

$$I_n' = \sqrt{2}\, I_n. \qquad (9\text{-}85)$$

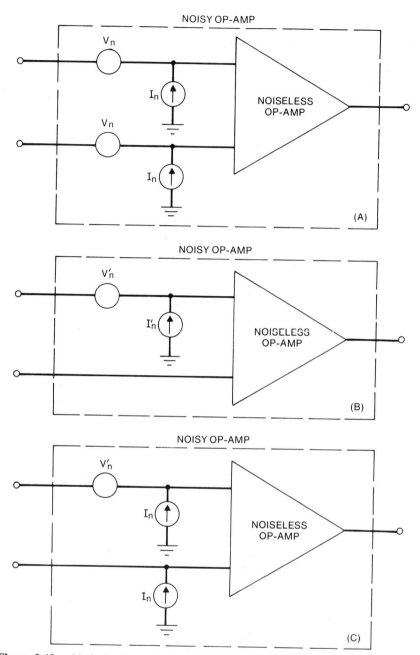

Figure 9-19. *Methods of modeling op-amp noise: (A) separate noise generators at each input; (B) noise generators combined at one input; (C) separate noise current generators with combined noise voltage generator.*

242

In still other cases, the noise voltage given by the manufacturer is the combined value V'_n, whereas the noise current is the value that applies to each input separately I_n. The equivalent circuit representing this arrangement is shown in Fig. 9-19C. The user, therefore, must be sure he understands which equivalent circuit is applicable to the data given by the manufacturer of the device before using the information. To date, there is no standard as to which of these three methods should be used for specifying op-amp noise.

Op-Amp Noise Factor

Normally, noise factor is not used in connection with op-amps. However, the noise factor can be determined by substituting Eq. 9-83 into Eq. 9-26, and solving for F. This gives

$$F = 2 + \frac{(V'_n)^2 + (I'_n R_s)^2}{4kTBR_s}. \tag{9-86}$$

Equation 9-86 assumes the source noise is due to the thermal noise in just one of the source resistors R_s, not both. This is a valid assumption when the op-amp is used as a single-ended amplifier. The thermal noise in the resistor R_s on the unused input is considered part of the amplifier noise and is a penalty paid for using this configuration.

In the case of the inverting op-amp configuration, the noise due to R_s at the unused input may be bypassed with a capacitor. This is not possible, however, in the noninverting configuration, since the feedback is connected to this point.

A second method of defining the noise factor for an op-amp is to assume the source noise is due to the thermal noise of both source resistors ($2R_s$ in this case). The noise factor then can be written as

$$F = 1 + \frac{(V'_n)^2 + (I'_n R_s)^2}{8kTBR_s}. \tag{9-87}$$

Equation 9-87 is applicable if the op-amp is used as a differential amplifier with both inputs driven.

SUMMARY

- If the source resistance is a variable and the source voltage a constant in the design of a circuit, minimizing noise factor does not necessarily produce optimum noise performance.
- For a given source resistance, the least noisy circuit is the one with the lowest noise factor.

- For the best noise performance the output signal-to-noise ratio should be maximized, this is equivalent to minimizing the total input noise voltage (V_{nt}).
- The concept of noise factor is meaningless when the source is a pure reactance.
- For best noise performance a low-source resistance should be used (assuming the source voltage remains constant).
- Noise performance may be improved by transformer coupling the source resistance to a value equal to $R_s = V_n/I_n$.
- If the gain of the first stage of a system is high, the total system noise is determined by the noise of the first stage.

BIBLIOGRAPHY

Baxandall, P. J., "Noise in Transistor Circuits," *Wireless World*, Vol. 74, November and December, 1968.

Cooke, H. F., "Transistor Noise Figure," *Solid State Design*, pp. 37–42, February, 1963.

Friis, H. T., "Noise Figures of Radio Receivers," *Proceedings of the IRE*, Vol. 32, July, 1944.

Gfeller, J., "FET Noise," *EEE*, pp. 60–63, June, 1965.

Graeme, J., "Don't Minimize Noise Figure," *Electronic Design*, January 21, 1971.

Haus, H. A. et al., "Representation of Noise in Linear Twoports," *Proceedings of IRE*, Vol. 48, January, 1960.

Letzter, S., and Webster, N., "Noise in Amplifiers," *IEEE Spectrum*, Vol. 7, No. 8, pp. 67–75, August, 1970.

Motchenbacher, C. D., and Fitchen, F. C., *Low-Noise Electronic Design*, Wiley, New York, 1973.

Mumford, W. W., and Scheibe, E. H., *Noise Performance Factors in Communication Systems*, Horizon House, Dedham, Mass. 1968.

Nielsen, E. G., "Behavior of Noise Figure in Junction Transistors," *Proceedings of the IRE*, Vol. 45, pp. 957–963, July, 1957.

Robe, T., "Taming Noise in IC Op-Amps," *Electronic Design*, Vol. 22, July 19, 1974.

Robinson, F. N. H., "Noise in Transistors," *Wireless World*, July, 1970.

Rothe, H., and Dahlke, W., "Theory of Noisy Fourpoles," *Proceedings of IRE*, Vol. 44, June, 1956.

Trinogga, L. A., and Oxford, D. F., "J.F.E.T. Noise Figure Measurement," *Electronic Engineering*, April, 1974.

Van der Ziel, A., "Noise in Solid State Devices and Lasers," *Proceedings of the IEEE*, Vol. 58, August, 1970.

Watson, F. B., "Find the Quietest JFETs," *Electronic Design*, November 8, 1974.

APPENDIX

A THE DECIBEL

One of the most commonly used, but often misunderstood terms in the field of electrical engineering is the decibel, abbreviated dB. The decibel is a logarithmic unit expressing the ratio of two powers. It is defined as

$$\text{Number of dB} = 10 \log \frac{P_2}{P_1}. \tag{A-1}$$

The unit can be used to express a power gain $(P_2 > P_1)$ or loss $(P_2 < P_1)$.

Since the definition of the decibel involves logarithms, it is appropriate to review some of their properties. The common logarithm Y of a number X is the power to which 10 must be raised to equal that number. Therefore, if

$$Y = \log X, \tag{A-2}$$

then

$$X = 10^Y. \tag{A-3}$$

Some useful identities involving logarithms are

$$\log AB = \log A + \log B, \tag{A-4}$$

$$\log \frac{A}{B} = \log A - \log B, \tag{A-5}$$

$$\log A^n = n \log A. \tag{A-6}$$

Using the Decibel for Other Than Power Ratios

It has become common practice to use dB to express voltage or current ratios. The commonly used definitions for voltage and current ratios expressed as dB are

$$\text{dB voltage gain} = 20 \log \frac{V_2}{V_1}, \tag{A-7}$$

$$\text{dB current gain} = 20 \log \frac{I_2}{I_1}. \tag{A-8}$$

245

These equations are only correct when both voltages, or both currents, are measured across equal impedances. Common usage, however, is to use the definitions in Eqs. A-7 and A-8 regardless of the impedance levels.

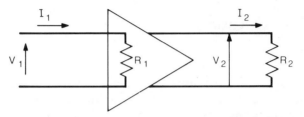

Figure A-1. *Circuit for comparing power gain and voltage gain.*

The relationship between voltage gain and power gain can be determined by referring to Fig. A-1. The power into the amplifier is

$$P_1 = \frac{V_1^2}{R_1}. \tag{A-9}$$

The power out of the amplifier is

$$P_2 = \frac{V_2^2}{R_2}. \tag{A-10}$$

The power gain G of the amplifier, expressed in dB is

$$G = 10 \log \frac{P_2}{P_1} = 10 \log \left(\frac{V_2}{V_1} \right)^2 \frac{R_1}{R_2}. \tag{A-11}$$

Using the identities of Eqs. A-4 and A-6, Eq. A-11 can be rewritten as

$$G = 20 \log \frac{V_2}{V_1} + 10 \log \frac{R_1}{R_2}. \tag{A-12}$$

Comparing Eq. A-12 with A-7 shows that the first term of the power gain is the voltage gain, as defined in Eq. A-7. If $R_1 = R_2$ then both the voltage gain and the power gain, expressed in dB, are numerically equal. The values of resistances R_1 and R_2 must be known, however, to determine the power gain from a given voltage gain.

In a similar manner, the power gain of the circuit in Fig. A-1 can be expressed as

$$G = 20 \log \frac{I_2}{I_1} + 10 \log \frac{R_2}{R_1}. \tag{A-13}$$

Notice that in this case the resistance ratio is the reciprocal of that in Eq. A-12.

Example A-1. A circuit has a voltage gain of 0.5, an input impedance of 100 Ω, and a load impedance of 10 Ω. From Eq. A-7, the dB voltage gain is −6 dB. From Eq. A-12

$$\text{dB power gain} = -6 + 10 \log \frac{100}{10} = 4 \text{ dB} \tag{A-14}$$

Therefore, in this case the power gain in dB is positive while the voltage gain in dB is negative.

Power Loss or Negative Power Gain

Let us compute the power gain from point 1 to point 2 for the case where the power at point 2 is less than the power at point 1. The power gain in dB is

$$G = 10 \log \frac{P_2}{P_1}. \tag{A-15}$$

To express the power ratio P_2/P_1 as a number greater than 1, we can rewrite Eq. A-15 as

$$G = 10 \log \left(\frac{P_1}{P_2} \right)^{-1}. \tag{A-16}$$

From the identity of Eq. A-6 this becomes

$$G = -10 \log \frac{P_1}{P_2}. \tag{A-17}$$

Therefore, power loss is indicated by a negative dB power gain.

Absolute Power Level

The decibel may also be used to represent an absolute power level by replacing the denominator of Eq. A-1 with a reference power P_0, such as 1 milliwatt. This gives

$$\text{Number of dB (absolute)} = 10 \log \frac{P}{P_0}, \tag{A-18}$$

and represents the absolute power level above or below the reference power. In this case, the user must know the reference power, which is expressed by adding additional letters to the abbreviation dB. For example, dBm is used to signify a reference power of 1 mW. Table A-1 lists some of the more commonly used dB units and their reference levels and abbreviations.

Table A-1   Reference Levels for Various dB Units

Unit	Type unit	Reference	Use	Remarks
dBa	Power	$10^{-11.5}$ W	Noise	Measured with F1A weighting
dBm	Power	1 mW		
dBrn	Power	10^{-12} W	Noise	
dBrnc	Power	10^{-12} W	Noise	Measured with "C-message" weighting
dBV	Voltage	1 V		
dBmV	Voltage	1 mV		
dBw	Power	1 W		
dBx	Power	(See remarks)	Crosstalk	90 dB of crosstalk coupling loss is the reference

Noise Measurements

In the telephone industry, noise on voice-frequency analog communications circuits is measured in terms of the annoyance effects of the noise on the listener. This is done by using a frequency weighting function, which accounts for the listeners' hearing as well as the frequency response of the telephone receiver. For example, if a 500-Hz interfering tone is deemed only half as annoying as a 1000-Hz interfering tone, the weighting function would assign only one half as much importance to the 500-Hz tone as to the 1000-Hz tone. The weighting function is physically obtained by placing an electric filter network in the noise meter.

Weighting Functions.   In the 1920s, the Western Electric 144-type telephone hand set was used for noise interference experiments. This resulted in the "144 weighting" curve shown in Fig. A-2. This curve is primarily determined by the frequency response of the 144-type hand set.

In the 1930s, the 302-type telephone set became prevalent and led to "F1A weighting." As shown in Fig. A-2, the F1A weighting has a wider bandwidth than 144 weighting. This is because the 302 telephone set itself had a wider response, and thus more noise could be transmitted through it to impair speech transmission.

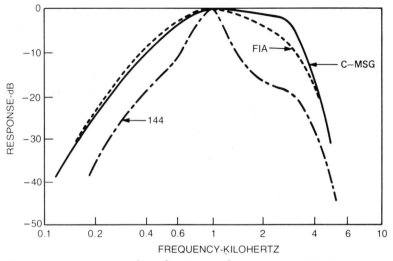

Figure A-2. *Various weighting functions used in noise measurements.*

With the advent of the 500-type telephone set in the 1950s, a new weighting function was developed. This has a slightly wider bandwidth and is known as "C-message weighting" (see Fig. A-2). C-message weighting is now the standard used for noise measurements in the Bell System.

Noise Units. When the early noise measuring sets were designed, it was decided to define noise in dB-type units compared to a reference noise power of 10^{-12} W, or -90 dBm. This amount of noise power (-90 dBm) is on the threshold of detection by the ear. The noise unit is called dBrn (dB reference noise). Thus, 0 dBrn means a noise power of -90 dBm. Such early test sets read 0 dBrn if -90 dBm of 1000-Hz power is measured. Due to the 144 weighting, however, equal amounts of power at other frequencies give different noise readings.

When the 302-type telephones became common, the 2B noise set incorporating F1A weighting was developed. The set's designer decided to make the 2B noise set give the same numerical reading as the early noise sets for measurements of 0- to 3000-Hz bandlimited white noise. Due to the different weighting networks used, however, the 2B noise set reads 5-dB lower than the early sets for 1000-Hz power. Thus, the reference power for the 2B set was raised to -85 dBm ($10^{-11.5}$ W) at 1000 Hz. The change in reference power necessitated a change in units. The new unit was called dBa (dBrn adjusted). Thus, 0 dBa is equal to -85 dBm of 1000-Hz power. This unit dBa, was used almost exclusively for 25 years.

When the 500-type telephone sets became widely used, the 3A noise measuring set was developed. This set incorporates "C-message weighting." It was decided to return to the original -90 dBm of 1000 Hz as the reference level, with dBrnc as the unit. The unit dBrnc means dBrn using C-message weighting. This reference level for the 3A noise measuring set was selected because the modern transmission circuits had become quieter, and it was thought that with a -85-db reference, negative dBa readings could occur and cause confusion. Because of the different reference levels, the 3A set reads 5-dB higher than the 2B set for 1000-Hz power. For most random noises, however, the 3A set reads about 6 dB higher than the 2B set due to the difference in weighting.

Table A-2 shows a comparison of readings from each of these three test sets.

Table A-2   Comparison of Noise Measurements Made with Various Weighting Functions

Western Electric noise test set	Weighting function	Test set reading for 0 dBm input	
		1000 Hz	0–3 kHz White noise
Early	144	90 dBrn	82 dBrn
2B	F1A	85 dBa	82 dBa
3A	C	90 dBrnc	88 dBrnc

Crosstalk Units

The unit of crosstalk is the dBx. This is an unusual unit since the reference is not an absolute power level. The reference is 90 dB loss from the interfering circuit to the circuit being interfered with. The unit is a measure of how much the crosstalk coupling loss is above 90 dB of coupling loss. Therefore,

$$dBx = 90 - (\text{crosstalk coupling loss in dB}) \qquad (A\text{-}19)$$

For example, suppose that circuit B picks up a signal from circuit A, but at a 62 dB lower power level. The crosstalk from A to B, then, is 28 dBx.

Summing Powers Expressed in Decibels

It is often necessary to determine the sum of two powers when the individual powers are expressed in dB with respect to some reference power level (e.g., dBm). The individual powers could always be converted to absolute power, added, and converted back to dB, but this is time consuming. The following procedure can be used when combining such terms.

Y_1 and Y_2 are two power levels expressed in dB above or below a reference power level P_0. P_1 and P_2 represent the absolute power levels corresponding to Y_1 and Y_2, respectively. Let us also assume that $P_2 \geq P_1$. From Eqs. A-18 and A-3 we can write

$$\frac{P_1}{P_0} = (10)^{Y_1/10} \qquad (\text{A-20})$$

and

$$\frac{P_2}{P_0} = (10)^{Y_2/10}, \qquad (\text{A-21})$$

therefore,

$$\frac{P_1}{P_2} = (10)^{(Y_1 - Y_2)/10}. \qquad (\text{A-22})$$

Let us define the difference (D) between the two powers, expressed in dB as

$$D = Y_2 - Y_1. \qquad (\text{A-23})$$

Therefore,

$$P_1 = P_2(10)^{-D/10}. \qquad (\text{A-24})$$

Adding P_2 to both sides gives

$$P_1 + P_2 = P_2(1 + 10^{-D/10}). \qquad (\text{A-25})$$

Expressing the sum of the powers P_1 and P_2 in terms of dB referenced to P_0 gives

$$Y_T = 10 \log\left(\frac{P_1 + P_2}{P_0}\right). \qquad (\text{A-26})$$

This can be rewritten as

$$Y_T = 10 \log(P_1 + P_2) - 10 \log P_0. \qquad (\text{A-27})$$

Substituting from Eq. A-25 for $P_1 + P_2$ gives

$$Y_T = 10 \log\left[P_2(1 + 10^{-D/10})\right] - 10 \log P_0 \qquad (\text{A-28})$$

or

$$Y_T = 10 \log\left(\frac{P_2}{P_0}\right) + 10 \log(1 + 10^{-D/10}). \qquad \text{(A-29)}$$

The first term represents Y_2, the larger of the two individual powers expressed in terms of dB. The second term represents how much Y_2 must be increased when the two are combined.

The sum of two powers expressed in dB is, therefore, equal to the larger power increased by

$$\boxed{10 \log(1 + 10^{-D/10})}$$

where D equals the difference in dB between the two original powers. The maximum value of this expression is 3 dB and occurs when $D = 0$. Values of this expression are tabulated in Table A-3.

Table A-3   Sum of Two Powers Expressed in dB

Amount D by which the two powers differ (dB)	Amount by which larger quantity should be increased to obtain the sum (dB)
0	3.00
0.5	2.77
1	2.54
1.5	2.32
2	2.12
3	1.76
4	1.46
5	1.19
6	0.97
7	0.79
8	0.64
9	0.51
10	0.41
11	0.33
12	0.27
15	0.14
20	0.04

B SUMMARY OF NOISE REDUCTION TECHNIQUES

The check list that follows is intended to summarize, in short form, the more commonly used noise reduction techniques. Those items with an asterisk are essentially free of added cost and should be used whenever applicable. The remaining techniques should be used whenever additional noise reduction is required.

Noise Reduction Check List

A. Suppressing Noise At Source:
- ☐ Enclose noise sources in a shielded enclosure.
- ☐ Filter all leads leaving a noisy environment.
- ☐ Limit pulse rise times.
- ☐ Relay coils should be provided with some form of surge damping.
- ☐ Twist noisy leads together.*
- ☐ Shield and twist noisy leads.
- ☐ Ground both ends of shields used to suppress radiated interference (shield does not need to be insulated).*

B. Eliminating Noise Coupling:
- ☐ Twist low-level signal leads.*
- ☐ Place low level leads near chassis (especially if the circuit impedance is high).
- ☐ Twist and shield signal leads (coaxial cable may be used at high frequencies).
- ☐ Shielded cables used to protect low-level signal leads should be grounded at *one end only* (coaxial cable may be used at high frequencies with shield grounded at both ends).*
- ☐ Insulate shield on signal-leads.
- ☐ When low-level signal leads and noisy leads are in the same connector, separate them and place the ground leads between them.*
- ☐ Carry shield on signal leads through connectors on a separate pin.
- ☐ Avoid common ground leads between high and low level equipment.*

253

☐ Keep hardware grounds separate from circuit grounds.*
☐ Keep ground leads as short as possible.*
☐ Use conductive coatings in place of nonconductive coatings for protection of metallic surfaces.
☐ Separate noisy and quiet leads.*
☐ Ground circuits at one point only (except at high frequencies).*
☐ Avoid questionable or accidental grounds.
☐ For very sensitive applications, operate source and load balanced to ground.
☐ Place sensitive equipment in shielded enclosures.
☐ Filter or decouple any leads entering enclosures containing sensitive equipment.
☐ Keep the length of sensitive leads as short as possible.*
☐ Keep the length of leads extending beyond cable shields as short as possible.*
☐ Use low-impedance power distribution lines.
☐ Avoid ground loops.*
☐ Consider using the following devices for breaking ground loops:

- Isolation transformers
- Neutralizing transformers
- Optical couplers
- Differential amplifiers
- Guarded amplifiers
- Balanced circuits.

C. Reducing Noise at Receiver:
☐ Use only necessary bandwidth.
☐ Use frequency selective filters when applicable.
☐ Provide proper power supply decoupling.
☐ Bypass electrolytic capacitors with small high-frequency capacitors.
☐ Separate signal, noisy, and hardware grounds.*
☐ Use shielded enclosures.
☐ With tubular capacitors, connect outside foil end to ground.*

*Essentially free of added cost.

C MULTIPLE REFLECTIONS OF MAGNETIC FIELDS IN THIN SHIELDS

Consider the case of a magnetic field with a wave impedance Z_1 incident on a thin shield of characteristic impedance Z_2, as shown in Chapter 6, Fig. 6-14. Since the shield is thin and the velocity of propagation is large, the phase shift through the shield can be neglected. Under these conditions, the total transmitted wave can be written as

$$H_{t(\text{total})} = H_{t2} + H_{t4} + H_{t6} + \cdots. \tag{C-1}$$

From Eqs. 6-10 and 6-15, we can write

$$H_{t2} = \frac{2Z_1 H_0}{Z_1 + Z_2} (e^{-t/\delta}) K, \tag{C-2}$$

where K equals the transmission coefficient at the second interface from medium 2 to medium 1 (Eq. 6-17).
    We can now write for H_{t4}

$$H_{t4} = \frac{2Z_1 H_0}{Z_1 + Z_2} (e^{-t/\delta})(1-K)(e^{-t/\delta})(1-K)(e^{-t/\delta})K, \tag{C-3}$$

which reduces to

$$H_{t4} = \frac{2Z_1 H_0}{Z_1 + Z_2} (e^{-3t/\delta})(K - 2K^2 + K^3). \tag{C-4}$$

Consider the case of a metallic shield where $Z_2 \ll Z_1$. Then $K \ll 1$ and $K^2 \ll K$ and $K^3 \ll K$, etc. The total transmitted wave can then be written as

$$H_{t(\text{total})} = 2H_0 K (e^{-t/\delta} + e^{-3t/\delta} + e^{-5t/\delta} + \cdots). \tag{C-5}$$

The infinite series in brackets in Eq. C-5 has the limit*

$$e^{-t/\delta} + e^{-3t/\delta} + e^{-5t/\delta} + \cdots = \frac{\operatorname{cosech}(t/\delta)}{2} = \frac{1}{2\sinh(t/\delta)} \tag{C-6}$$

*Standard Mathematical Tables; 21st edition; p. 343; Chemical Rubber Co., 1973.

Substituting Eq. 6-17 for K and Eq. C-6 for the infinite series in Eq. C-5, gives

$$H_{t(total)} = \frac{4H_0Z_2}{Z_1}\left[\frac{1}{2\sinh(t/\delta)}\right], \tag{C-7}$$

or

$$\frac{H_0}{H_{t(total)}} = \left(\frac{Z_1}{4Z_2}\right)2\sinh\left(\frac{t}{\delta}\right). \tag{C-8}$$

Shielding effectiveness is 20 times the log of Eq. C-8 or

$$S = 20\log\frac{Z_1}{4Z_2} + 20\log\left[2\sinh\left(\frac{t}{\delta}\right)\right]. \tag{C-9}$$

Replacing Z_1 with the impedance of the wave at the shield Z_w and replacing Z_2 with the shield impedance Z_s, gives

$$S = 20\log\frac{Z_w}{4Z_s} + 20\log\left[2\sinh\left(\frac{t}{\delta}\right)\right]. \tag{C-10}$$

The first term of Eq. C-10 is the reflection loss R, as defined by Eq. 6-22. To calculate the correction factor B we must put Eq. C-10 into the form of Eq. 6-3. The second term of Eq. C-10 must therefore be equal to $A + B$. Thus, we can write

$$B = 20\log\left[2\sinh\left(\frac{t}{\delta}\right)\right] - A. \tag{C-11}$$

Substituting for A, from Equation 6-12a, gives

$$B = 20\log\left(2\sinh\frac{t}{\delta}\right) - 20\log e^{t/\delta}. \tag{C-12}$$

Combining terms

$$B = 20\log\left[\frac{2\sinh\left(\frac{t}{\delta}\right)}{e^{t/\delta}}\right], \tag{C-13}$$

expressing the sinh (t/δ) as an exponential, gives the correction factor B as

$$B = 20\log[1 - e^{-2t/\delta}]. \tag{C-14}$$

Figure 6-15 is a plot of Eq. C-14 as a function of t/δ. Note that the correction factor B is always a negative number, indicating that less shielding is obtained from a thin shield due to the multiple reflections.

Table C-1 lists values of B for very small values of t/δ which are not shown in Fig. 6-15.

Table C-1   Reflection Loss Correction Factor (B) for Very Thin Shields

t/δ	B(dB)
0.001	-54
0.002	-48
0.004	-42
0.006	-38
0.008	-36
0.01	-34
0.05	-20

APPENDIX

D PROBLEMS

PROBLEM 1-1

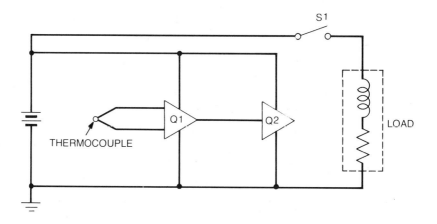

In the circuit shown, amplifiers $Q1$ and $Q2$, are used to amplify the low level signal from a thermocouple. A high-power load, periodically operated by switch $S1$, is also powered from the same battery. Assuming the circuit is wired as shown in the figure, identify potential noise sources, coupling channels, and receivers.

PROBLEM 2-1

The stray capacitance between conductors 1 and 2 is 50 pF. Each conductor has a capacitance to ground of 150 pF. Conductor 1 has a 10-V ac signal at a frequency of 100 kHz on it. See figure at top of page 259. What is the noise voltage picked up by conductor 2 if it is terminated (R_T):

  a. in an infinite resistance?

  b. in a 1000-Ω resistance?

  c. in a 50-Ω resistance?

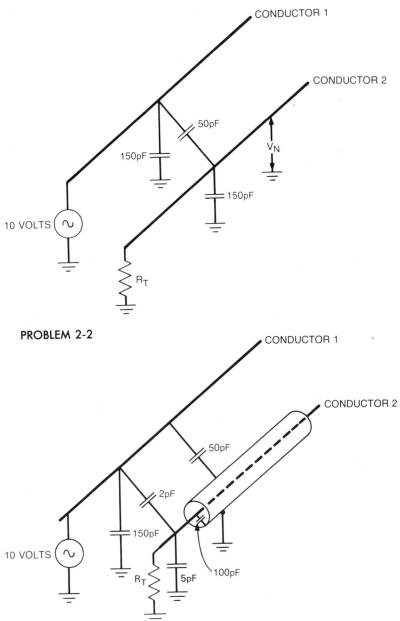

PROBLEM 2-2

In the above illustration, a grounded shield is placed around conductor 2. The capacitance from conductor 2 to the shield is 100 pF. The capacitance between conductors 2 and 1 is 2 pF, and the capacitance between conductor 2 and ground is 5 pF. Conductor 1 has a 10-V ac signal at a frequency of 100 kHz on it. For this configuration, what is the noise voltage picked up by conductor 2 if it is terminated (R_T):

   a. in an infinite resistance?

   b. in a 1000-Ω resistance?

   c. in a 50-Ω resistance?

PROBLEM 2-3

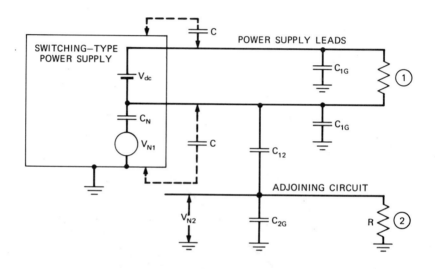

Due to the switching action of power transistors, a noise voltage is usually introduced in switching-type power supplies between the power supply output leads and the case. This is V_{N1} in the illustration. This noise voltage can capacitively couple into adjoining circuit 2 as illustrated. C_N is the equivalent coupling capacitance between the case and the output power leads.

   a. For the above circuit configuration, determine and sketch the ratio V_{N2}/V_{N1} as a function of frequency. (Neglect the capacitors C, shown dotted.)

Next, capacitors (C) are added between the output power leads and the case, as indicated.

   b. How does this effect the noise coupling?

   c. How would shielding of the power-supply leads improve the noise performance?

PROBLEM 2-4

Two conductors, each 10-cm long and spaced 1-cm apart, form a circuit. This circuit is located where there is a 10-gauss magnetic field at 60 Hz.

What is the maximum noise voltage coupled into the circuit due to the magnetic field?

PROBLEM 2-5

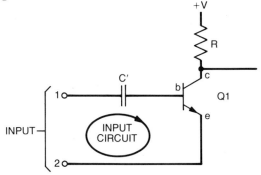

A

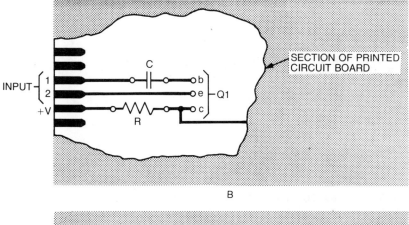

B

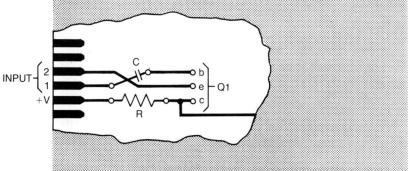

C

Figure A in the illustration is a partial schematic for a low-level transistor amplifier. A printed circuit layout for the circuit is shown in Fig. B. The circuit is located within a strong magnetic field.

What is the advantage of the alternate layout shown in Fig. C over that of Fig. B?

PROBLEM 2-6

For the case of magnetic field coupling, calculate the shield factor (η) for a shielded (coaxial) receiver circuit (shield grounded at both ends) in terms of the resistance and inductance of the shield.

PROBLEM 3-1

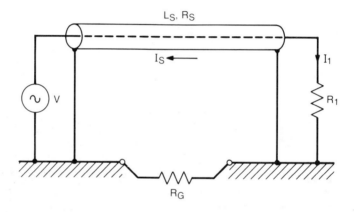

In the illustration, the shield is characterized by its inductance L_S and resistance R_S. An equivalent ground resistance R_G is also associated with the above arrangement.

a. Make an asymptotic plot of $|I_S/I_1|$ versus frequency.

b. Above what frequency does 98% of the current I_1 return through the shield?

PROBLEM 3-2

If a bead of magnetic material is placed on a shielded cable, what effect does this have on the shield cutoff frequency?

PROBLEM 3-3

A magnetic field induces a noise voltage into the following circuit.

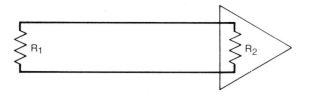

a. What is the noise voltage at the amplifier input terminals as a function of R_1?

b. How do you explain the answer to part "a" when compared to the statement in the text that the impedance of the receiver circuit does not affect the magnetic pickup?

PROBLEM 3-4

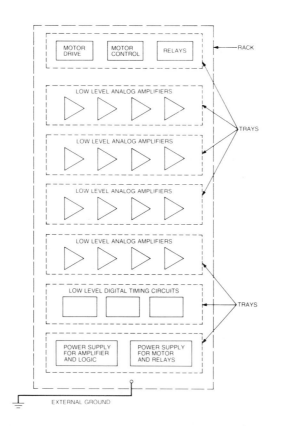

For the physical arrangement of a tape recorder shown in the above illustration, design a grounding system.

PROBLEM 3-5

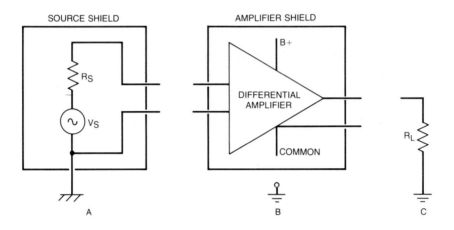

Determine the optimum cabling and grounding arrangement for the circuit illustrated. The circuit consists of a grounded, low-level, low-frequency, signal source at location A; a differential amplifier at location B; and a grounded load at location C. Do not use any transformers or guard shields. The source at A and the load at C must remain grounded.

PROBLEM 3-6

A longitudinal choke (neutralizing transformer) is placed in series with a transmission line connecting a low-level source to a 900-Ω load. The transmission-line conductors each have a resistance of 1 Ω. Each winding of the longitudinal choke has an inductance of 0.044 H and a resistance of 4 Ω.

a. Above what frequency will the choke have a negligible effect on the signal transmission?

b. How much attenuation does the choke provide to a ground differential noise voltage at 60, 180, and 300 Hz?

PROBLEM 3-7

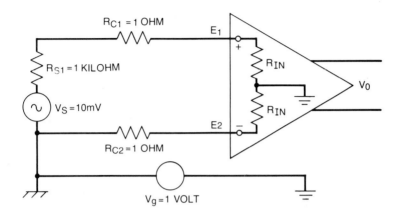

For the circuit illustrated, what restriction must there be on the value of R_{IN} in order to keep the ground noise coupled into the differential amplifier to less than 0.1% of the signal voltage (V_s)?

PROBLEM 3-8

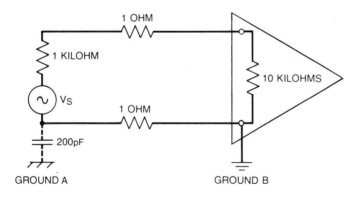

The circuit in the illustration has 200 pF of parasitic capacitance from source to ground. What is the noise voltage at the amplifier if the noise voltage between the two grounds is:

a. 100 mV at 60 Hz?
b. 100 mV at 6000 Hz?

PROBLEM 3-9

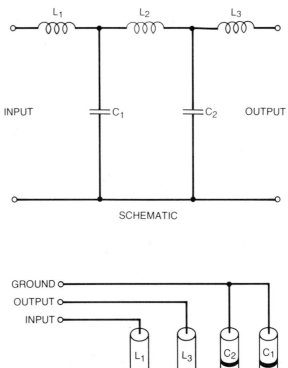

SCHEMATIC

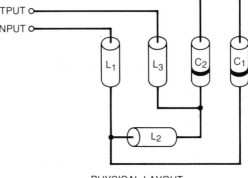

PHYSICAL LAYOUT

The illustration shows the schematic and a physical layout for a high-frequency, low-pass filter. The inductors are wound as solenoids on open magnetic cores. The capacitors are tubular.

   a. List the disadvantages of the layout shown.

   b. Propose a new layout which overcomes these disadvantages.

PROBLEM 3-10

A digital voltmeter with guard shield is used to make a voltage measurement in the following bridge circuit.

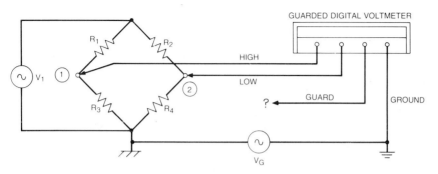

a. What are the common-mode sources?

b. Where should the guard shield be connected and why?

Note: The correct solution to this problem requires careful thought and analysis.

Hint: Take the Thevenin's equivalent circuit looking into terminals 1 and 2.

PROBLEM 4-1

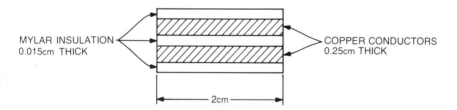

MYLAR INSULATION 0.015cm THICK

COPPER CONDUCTORS 0.25cm THICK

2cm

The power bus arrangement shown in the illustration is used to transmit 5 V dc to a 10-A load. The bus bar is 5 m long.

a. What is the dc voltage drop in the distribution system?

b. What is the characteristic impedance of the line?

PROBLEM 5-1

Power is supplied to a two-stage high-frequency electronic circuit by a 20-gauge wire of length $l(l \gg 2'')$ and via two feed-through capacitors, as illustrated. Stage 2 of the circuit operates at a frequency of 25 MHz, and each stage is in a shielded enclosure to prevent coupling between the stages. In this arrangement, a resonant circuit is formed by the power line inductance and the two feed-through capacitors. This resonant circuit could possibly provide a coupling path between the two circuits.

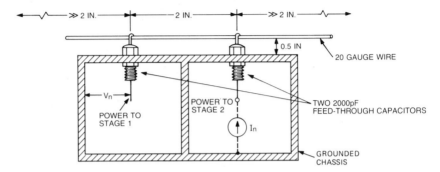

a. What is the resonant frequency and damping factor of this L–C circuit? (Use wire data given in Table 5-3 on p. 128.)

b. If the noise source can be represented by an equivalent current source $I_n(j\omega)$ in stage 2, as shown in the illustration, derive an expression for the noise voltage $V_n(j\omega)$ coupled into the first stage.

c. For $I_n = 1$ mA and $f = 25$ MHz, determine $|V_n|$.

d. Propose a way to reduce noise coupling between stages 1 and 2 by using ferrite beads with the characteristics given in Fig. 5-16 of the text. Consider both beads 1 and 2. What effect does the bead have on the dc voltage of both stages?

e. Repeat part "c" of this problem for your proposed, improved arrangement.

PROBLEM 5-2

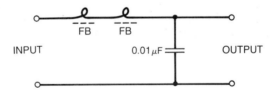

What is the attenuation at 5 MHz of a filter consisting of two ferrite beads (bead 2 in Fig. 5-16) and a 0.01-μF capacitor connected as shown.

PROBLEM 5-3

Make a table of the ratio of the ac resistance to the dc resistance of a 22-gauge copper conductor at the following frequencies: 0.2, 0.5, 1, 2, 5, 10, and 50 MHz.

PROBLEM 5-4

A copper conductor has a rectangular cross section of 0.5×2 cm.

a. Calculate the dc resistance per 100 ft of conductor length.

b. Calculate the 1-MHz resistance per 100 ft of conductor length.

PROBLEM 6-1

What is the magnitude of the characteristic impedance of silver, brass, and stainless steel at 10 kHz?

PROBLEM 6-2

Calculate the skin depth and absorption loss for a brass shield 0.062-in. thick at the following frequencies:

a. 0.1 kHz.

b. 1.0 kHz.

c. 10 kHz.

d. 100 kHz.

PROBLEM 6-3

Considering absorption loss only, discuss the design of a shield to provide 30 dB of attenuation against a 60-Hz field.

PROBLEM 6-4

a. What is the reflection loss of a 0.001-in. thick copper shield to a 1000-Hz electric field?

b. If the thickness is increased to 0.01 in., what is the reflection loss?

PROBLEM 6-5

Calculate the shielding effectiveness of a 0.015-in. thick copper shield located 2.5 cm from the source of a 10-kHz magnetic field.

PROBLEM 6-6

What would be the shielding effectiveness of the shield of the previous problem if it were located in the far field?

PROBLEM 6-7

What is the shielding effectiveness of a 0.032-in. thick aluminum shield located 1 ft away from the source of a 10-kHz electric field?

PROBLEM 6-8

A shield is located 6 in. from the source of an electric or magnetic field. Above what frequency should the far-field equations be used?

PROBLEM 6-9

Calculate the absorption loss of three different copper shields, 0.020-in. 0.040-in. and 0.060-in. thick, to a 1-kHz magnetic field.

PROBLEM 6-10

List as many reasons as you can why a "paint can" would make a good medium-to-high frequency shield.

PROBLEM 7-1

A 1-H, 400-Ω relay coil is to be operated from a 30-V dc supply. The switch controlling the relay has platinum contacts. Design a contact-protection network for this circuit.

PROBLEM 7-2

For the zener diode protection circuit of Fig. 7-15E, plot the following three waveshapes when the contact closes and then opens. Assume no contact breakdown.

   a. The voltage across the load (V_L).
   b. The current through the load (I_L).
   c. The voltage across the contact (V_c).

PROBLEM 8-1

Calculate the noise voltage produced by a 5000-Ω resistor in a system with a 10-kHz bandwidth, at a temperature of:

   a. 27°C (300°K).
   b. 100°C (373°K).

PROBLEM 8-2

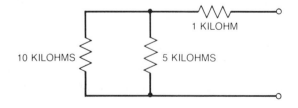

Calculate the thermal noise voltage per square root of bandwidth for the circuit at the bottom of page 270, at room temperature.

PROBLEM 8-3

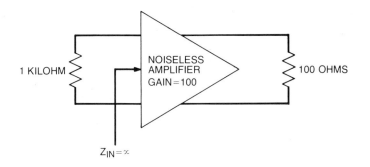

Determine the noise voltage at the amplifier output for the circuit illustrated. Assume the amplifier has a frequency response equivalent to:

a. An ideal low pass filter with a cutoff frequency of 2 kHz.
b. An ideal bandpass filter with cutoff frequencies of 99 and 101 kHz.

PROBLEM 8-4

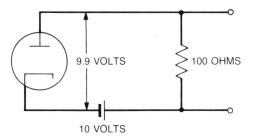

What is the total noise voltage across the output terminals of the circuit illustrated? Calculate the effect of shot noise as well as thermal noise for a bandwidth of 2.5 kHz. The diode is operated in the temperature-limited region.

PROBLEM 8-5

Determine the noise voltage per square root of bandwidth generated across terminals $A-A$ for the circuit at the top of page 272, at room temperature and at a frequency of 1590 Hz.

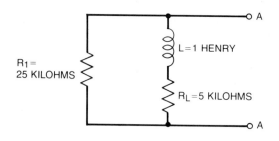

PROBLEM 9-1

Derive Eq. 9-3 (in text) from Eq. 9-1.

PROBLEM 9-2

Which device produces the least equivalent input device noise ($V_{nd}/\sqrt{B}$)?

   a. A bipolar transistor with a noise figure of 10 dB measured at $R_s = 10^4\ \Omega$.

   b. An FET with a noise figure of 6 dB measured at $R_s = 10^5$-Ω.

PROBLEM 9-3

A transistor has a noise figure of 3 dB measured with a source resistance of 1.0 MΩ. What is the output-power signal-to-noise ratio if this transistor is used in a circuit with an input signal of 0.1 mV and a source resistance of 1.0 MΩ? Assume the system has an equivalent noise bandwidth of 10 kHz.

PROBLEM 9-4

The noise of an FET is specified as follows. Equivalent input noise voltage is 0.06×10^{-6} V/$\sqrt{\text{Hz}}$, and the equivalent input noise current is 0.2×10^{-12} A/$\sqrt{\text{Hz}}$.

   a. If the FET is used in a circuit with a source resistance of 100-Ω and an equivalent noise bandwidth of 10 kHz, what is the noise figure?

   b. What value of R_s will produce the lowest noise figure and what is the noise figure with this value of R_s?

PROBLEM 9-5

A low-noise preamplifier is to be driven from a 10-Ω source. Data supplied by the manufacturer specifies V_n and I_n at the operating frequency as:

$$\frac{V_n}{\sqrt{B}} = 10^{-8} V/\sqrt{Hz} ,$$

$$\frac{I_n}{\sqrt{B}} = 10^{-13} A/\sqrt{Hz} .$$

a. Determine the input-transformer turns ratio to provide optimum noise performance.

b. Calculate the noise figure for the circuit using the transformer of part "a."

c. What would be the noise figure with the preamplifier directly coupled to the 10-Ω source?

d. What would be the signal to noise improvement factor (SNI) for this circuit?

PROBLEM 9-6

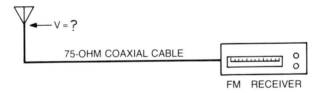

The illustration shows an FM antenna connected to an FM receiver by a section of 75-Ω matched coaxial cable. The required signal-to-noise ratio at the input terminals to the set for good quality reception is 18 dB, and the noise figure of the receiver is 8 dB.

a. If the cable connecting the receiver to the antenna has 6 dB of insertion loss, what signal voltage is required at the point where the antenna connects to the cable, to provide good quality reception? The noise bandwidth of the receiver is 50 kHz.

b. Why is the voltage considerably less than that required in the case of a TV set as worked out on p. 230 of the text?

PROBLEM 9-7

Find the noise figure for a system with a noise temperature (T_e) equal to 290°K.

PROBLEM 9-8

A transistor is operated at a frequency $f \ll f_\alpha$. The transistor parameters are $r_b' = 50 \ \Omega$ and $\beta_0 = 100$. Calculate the minimum noise factor and the value of source resistance for which it occurs when collector current is:

   a. $10 \ \mu A$.
   b. $1.0 \ m/A$.

Note: $r_e \approx 26/I_C(ma)$

PROBLEM 9-9

A junction FET has the following parameters measured at 100 MHz: $g_{fs} = 1500 \times 10^{-6}$ mhos and $g_{11} = 800 \times 10^{-6}$ mhos. If the transistor is to be used in a circuit with a source resistance of 1000 Ω, what is the noise figure?

PROBLEM 9-10

Derive Eq. 9-37. Start with the equivalent circuit of Fig. 9-4 and Eq. 9-1.

PROBLEM 9-11

Derive Eq. 9-40. Start with the equivalent circuit of Fig. 9-4.

E ANSWERS TO PROBLEMS

Note: For some of these problems, a single unique solution may not exist. Therefore, solutions other than the ones listed here may also be acceptable.

PROBLEM 1-1

SOURCE	COUPLING CHANNEL	RECEIVER
Arc in Switch	Radiation	TC* Q1 Q2
	Radiation + Conduction	Q1 Q2
Transient Load Current	Radiation	TC* Q1 Q2
	Common Battery Resistance + Conduction	Q1 Q2
	Common Impedance of Ground	Q1 Q2
	Radiation + Conduction	Q1 Q2
Steady State Load Current	Common Impedance of Ground	Q1 Q2
Magnetic Field of Inductor	Radiation	TC* Q1 Q2
	Radiation + Conduction	Q1 Q2

*Thermocouple

PROBLEM 2-1

a. 2.5 V.
b. 314 mV.
c. 15.7 mV.

PROBLEM 2-2

a. 187 mV.
b. 12.6 mV.
c. 628 μV.

PROBLEM 2-3

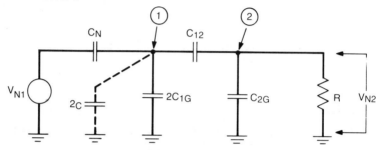

An equivalent circuit of the noise coupling is shown above. A reasonable assumption to simplify the problem is that $2C_{1G} \gg C_{12}$.

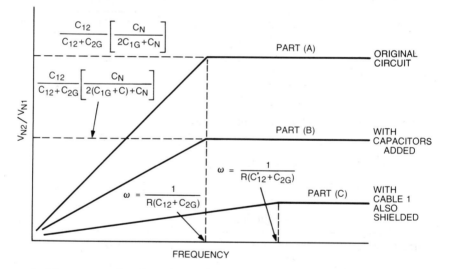

a. The asymtotic plot of V_{N2}/V_{N1} is shown as the top curve in the above plot.

b. If capacitors C are added, it has the effect of increasing C_{1G}, which decreases the maximum coupling while keeping the break point constant. This is shown as the middle curve in the above plot.

c. Shielding cable 1 reduces C_{12} to a value C'_{12}. This lowers the maximum coupling further and also increases the break frequency. This is shown as the bottom curve in the plot. A second effect of shielding is to increase capacitance C_{1G}, which further decreases the coupling.

PROBLEM 2-4

376 μV.

PROBLEM 2-5

The magnetic field pickup is of opposite polarity on either side of the crossover, thereby, canceling the noise voltage.

PROBLEM 2-6

$$\eta = \frac{R_S/L_S}{j\omega + R_S/L_S} \; .$$

Hint: from Eq. 2-17, $M_{23} = L_S$.

PROBLEM 3-1

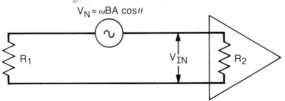

FREQUENCY

a. This solution is shown in the illustration above.

b. $f_{98\%} = [5(R_G + R_S)]/2\pi L_S$.

PROBLEM 3-2

The addition of magnetic material increases the shield inductance and hence decreases the shield cutoff frequency.

PROBLEM 3-3

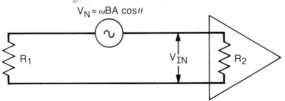

For magnetic coupling, we have the equivalent circuit shown at the bottom of page 277.

a. $V_{IN} = V_N R_2 / (R_1 + R_2)$

b. The total voltage (V_N) magnetically coupled into the circuit is not a function of the values of resistors R_1 and R_2. However, how this voltage distributes itself between resistors R_1 and R_2 is a function of their relative values.

PROBLEM 3-4

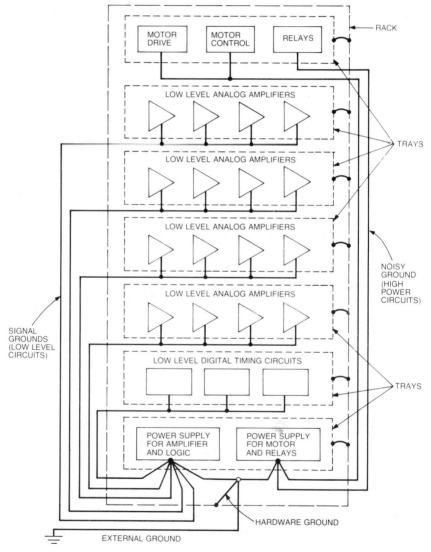

PROBLEM 3-5

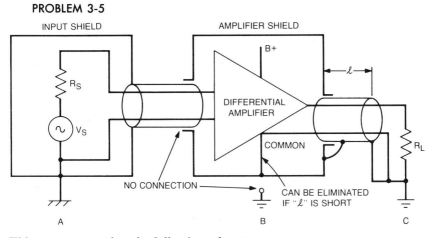

This arrangement has the following advantages.

- Input shield grounded at source.
- Only one ground on input shield.
- Amplifier shield connected to amplifier common.
- Output shield connected to load ground.
- Only one ground on output shield.
- Protection against ground noise differential is obtained by large input impedance of amplifier.

PROBLEM 3-6

a. 90.4 Hz.

b. 10.8 dB at 60 Hz.
   20 dB at 180 Hz.
   24 dB at 300 Hz.

PROBLEM 3-7

$R_{IN} \geq 100$ MΩ.

PROBLEM 3-8

a. 6.85×10^{-9} V.

b. 6.85×10^{-7} V.

PROBLEM 3-9

Question a.

- Maximum coupling exists between L_1 and L_3 since they are close together and parallel.
- Large capacitance exists between input and output leads.
- Ground lead is common to both the input and the output.
- Outside foil (banded) end of capacitors C_1 and C_2 not grounded. (See Fig. 5-4.)
- Stray capacitance across L_2 is increased due to conductor between L_1 and C_1 and the conductor between L_3 and C_2 being close to and parallel to L_2.
- Long lead connecting L_1 to C_1 increases the inductance in series with C_1 thus lowering its self-resonant frequency.

Question b.

See figure below.

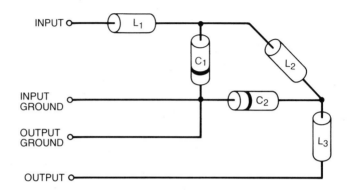

PROBLEM 3-10

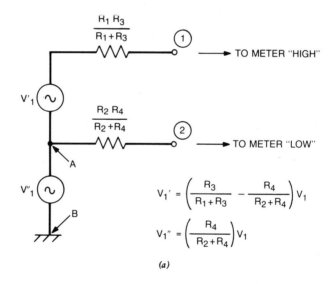

$$V_1' = \left(\frac{R_3}{R_1+R_3} - \frac{R_4}{R_2+R_4}\right)V_1$$

$$V_1'' = \left(\frac{R_4}{R_2+R_4}\right)V_1$$

(a)

a. V_G and that portion of V_1 that appears across $R_4(V_1'')$.

b. Thevenin's equivalent circuit looking into terminals 1 and 2 is shown above. The ideal guard connection is to point A. Point A, however, does not exist in the actual circuit. There are, therefore, two possible alternatives.

1. Connect guard shield to point B.

2. A better solution, if required, is to generate a new point at the same potential as point A, then connect the guard to this point. This is shown below. R_5 and R_6 must satisfy the following:

$$\frac{R_6}{R_5+R_6}=\frac{R_4}{R_4+R_2}, \quad \text{and} \quad R_5+R_6 \ll R_4+R_2.$$

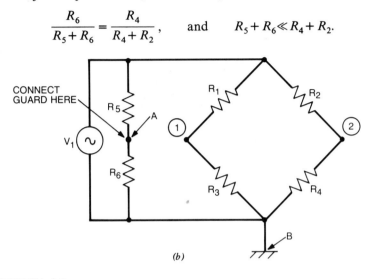

(b)

PROBLEM 4-1

a. 34.5 mV.

b. 1.26 Ω.

PROBLEM 5-1

a. $\delta = 1.31 \times 10^{-4}$,
   $f_r = 25$ MHz.

b.

$$V_n = \frac{I_n}{2j\omega C\left[(j\omega)^2 + LC/2 + j\omega RC/2 + 1\right]}.$$

c. $V_n = 43.8$ mV.

d. Use ferrite bead number 1 on the power line between the two feed through capacitors.

e. Using bead 1, which represents a 75-Ω resistor at 25 MHz.

$$V_n = 0.14 \text{ mV}.$$

PROBLEM 5-2

21.2 dB.

PROBLEM 5-3

For $d = 0.0253$ in.

Frequency (MHz)	R_{ac}/R_{dc}
0.2	1.35
0.5	1.98
1	2.69
2	3.70
5	5.70
10	7.95
50	17.47

PROBLEM 5-4

a. $R_{dc} = 5.25 \times 10^{-3}$ Ω per 100 ft.

b. $R_{1MHz} = 0.317$ Ω per 100 ft.

PROBLEM 6-1

Silver: $|Z_s| = 3.6 \times 10^{-5}$ Ω.

Brass: $|Z_s| = 7.2 \times 10^{-5}$ Ω.

Stainless: $|Z_s| = 5.8 \times 10^{-3}$ Ω.

PROBLEM 6-2

Frequency (kHz)	Skin Depth (in.)	Absorption Loss (dB)
0.1	0.51	1.1
1.0	0.16	3.3
10	0.05	10.6
100	0.02	33.4

PROBLEM 6-3

Using nonferrous material, this would require a shield thickness of 1.2 in. or more; this is impractical.

Using steel, however, the shield would have to be 0.12 in. thick; this is considerably more reasonable. A high permeability material such as mu-metal could also be used—in this case the required thickness would be less than the 0.12 in. required for steel.

PROBLEM 6-4

a. 138 dB.
b. 138 dB.

PROBLEM 6-5

24 dB.

PROBLEM 6-6

133 dB.

PROBLEM 6-7

134 dB.

PROBLEM 6-8

Greater than 313 MHz.

PROBLEM 6-9

Thickness (in.)	Absorption Loss (dB)
0.020	2.11
0.040	4.23
0.060	6.34

PROBLEM 6-10

- Steel walls—good absorption loss.
- Tin plate—increases reflection loss.
- Multiple Layer Shield (tin–steel–tin).
- Soldered seams.
- Pressure contact lid.

PROBLEM 7-1

270 Ω in series with 0.1 μF across the load or the contact.

PROBLEM 7-2

Approximate waveshapes are shown below.

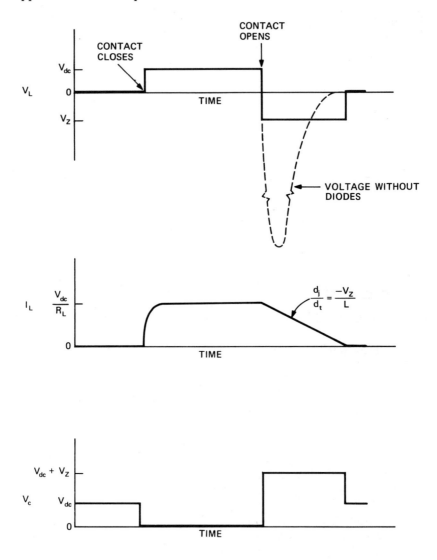

PROBLEM 8-1

a. 0.91×10^{-6} V.
b. 1.01×10^{-6} V.

PROBLEM 8-2

8.33×10^{-9} V$/\sqrt{\text{Hz}}$.

PROBLEM 8-3

a. 179 μV.
b. 179 μV.

PROBLEM 8-4

110×10^{-9} V.

PROBLEM 8-5

10×10^{-9} V$/\sqrt{\text{Hz}}$.

PROBLEM 9-2

- Bipolar: 3.8×10^{-8} V$/\sqrt{\text{Hz}}$.
- FET: 7×10^{-8} V$/\sqrt{\text{Hz}}$.

Therefore, the bipolar transistor produces the least equivalent input device noise.

PROBLEM 9-3

- 14.9 dB.

PROBLEM 9-4

a. $NF = 5.4$ dB.
b. $NF = 4.0$ dB.

PROBLEM 9-5

a. Turns Ratio $= 100$.
b. $NF = 0.5$ dB.
c. $NF = 27.9$ dB.
d. $SNI = 556$.

PROBLEM 9-6

a. 5 μV.
b. The inherent noise immunity of FM allows operation at a lower signal-to-noise ratio, and the 75-Ω transmission line has less thermal noise

than a 300-Ω system. Also, the smaller bandwidth allows less noise into the system.

PROBLEM 9-7

3 dB.

PROBLEM 9-8

a. $F = 1.11$, $R_s = 26,500\ \Omega$.
b. $F = 1.25$, $R_s = 572\ \Omega$.

PROBLEM 9-9

$NF = 6$ dB.

INDEX